THE
BIOCHEMIC
HANDBOOK

P9-CJF-083

Revised 1976

Thirty - third printing since 1970

Originally published as
BIOCHEMIC THEORY AND PRACTICE
by J. B. Chapman, M.D. and
Edward L. Perry, M.D.

ISBN 0-89378-051-0

Formur, Incorporated—Publishers

St. Louis, Missouri 63108

The Biochemic Theory

THE Biochemic System of Medicine is based primarily and fundamentally upon the "cell theory" of Virchow. In 1858 that great scientist pronounced the now famous dictum that the body is merely a collection of cells, and that medicinal treatment should be directed towards the individual cell. This great truth once enunciated, was seized upon, developed and elaborated by others, notably Moleschott of Rome, and Schuessler of Oldenburg, until with the full appreciation of the value of the inorganic constituents of the cell substance, and the part taken by them in the preservation of the health of the human organism, the Biochemic treatment of disease truly became a system of medicine.

In this system for the first time the paramount importance of the inorganic constituents of the cell substance was recognized, the fact being that these "cell salts," as they are commonly called, are the vital portions of the body, the workers, the builders; that the water and organic substances forming the remainder of the organism are simply inert matter used by these salts in building the cells of the body.

Should a deficiency occur in one or more of these workers, of whom there are twelve, some abnormal condition arises. These abnormal conditions are known by the general term disease, and according as they manifest themselves in different parts of the body, they have been designated by various names. But these names totally fail to express the real

1

trouble. Every disease which afflicts the human race is due to a lack of one or more of these inorganic workers. Every pain or unpleasant sensation indicates a lack of some inorganic constituent of the body. Health and strength can be maintained only so long as the system is properly supplied with these cell-salts.

These cell salts are twelve in number, all essential to the proper growth and development of every part of the body. They are the:

Phosphates of
Lime, Calcarea Phosphoricum.
Iron, Ferrum Phosphoricum.
Potash, Kali Phosphoricum.
Soda, Natrum Phosphoricum.
Magnesia, Magnesia Phosphoricum

Chlorides of
Potash, Kali Muriaticum.
Soda, Natrum Muriaticum.

Sulphates of
Lime, Calcarea Sulphurica.
Soda, Natrum Sulphuricum.
Potash, Kali Sulphuricum.

Fluoride and Pure Silica
of Lime, Calcarea Fluorica,
Silicea.

The inorganic materials of nerve cells are Magnesia Phos., Kali Phos., Natrum and Ferrum. Muscle cells contain the same, with the addition of Kali Mur. Connective tissue cells have for their specific substance Silicea, while that of the elastic tissue cells is Calcarea Fluor. In bone cells we have Calcarea Fluor., and Magnesia Phos., and a large portion of Calcarea Phos. This latter is found in small quantities in the cells of muscle, nerve, brain

2

and connective tissue. Cartilage and mucous cells have for their specific inorganic material Natrum Mur., which occurs also in all solid and fluid parts of the body. Hair and the crystalline lens contain, among other inorganic substances, also Ferrum. The carbonates, as such, are, according to Moleschott, without any influence in the process of cell formation.

They are reserve materials out of which phosphates and sulphates can form themselves.

Sulphuric and phosphoric acids unite with the bases of the carbonates and carbonic acid is given off. In this way sulphates and phosphates are produced.

The oxygen of the air, upon reaching the tissues through the blood by means of the respiration, acts upon the organic substances which are to enter into the formation of new cells. The products of this action are the organic materials which form the physical basis of muscle, nerve, connective tissue and mucous substances; each of these substances is the basis of a particular group of cells, to which by means of chemical affinity, the above mentioned cell salts are united, and then new cells are produced. With the production of new cells there occurs at the same time a destruction of the old ones, resulting from the action of oxygen on the organic substances forming the basis of these cells. This oxidation has, as a consequence, a breaking-down effect on the cells themselves.

Urea, uric acid and sulphuric acid are the result of the oxidation of the albuminous substances, while phosphoric acid is produced by the oxidation of lecithine contained in the nervous tissue, brain, spinal cord and blood corpuscles. Lactic acid results from the fermentation of milk-sugar, and finally breaks down into carbonic acid and water. Sulphuric and phosphoric acids unite with the basis of the carbonates, forming sulphates and phosphates, and set free carbonic acid.

Uric acid unites with sodium, forming sodium urate, which, being of no use to the animal economy is eliminated from the system; while partial failure of this, and its accumulation in the neighborhood of joints, together with albuminous substances, gives rise to gout. Natrum Sulph. removes the water resulting from the oxidation of the organic substances of the body, in which are suspended or dissolved the mineral matters set free in the retrograde cell metamorphosis, as well as the newly formed organic substances, such as urea, uric acid, etc.

Disturbance of the function of the molecules of Natrum Sulph., may be followed, according to its duration or extent, as well as its location, by a retarded removal of this water of oxidation and its dissolved or suspended matter. This implies a slower tissue change and a consequent liability to diabetes, gout, etc.

It is interesting to note that Natrum Sulph. and Natrum Mur. act in opposite ways; for while the former—the sulphate—removes from the tissues the water, according to the process just described, the muriate—the common salt—enters the tissues dissolved in the water from the blood plasma, in order that the requisite degree of moisture proper for each tissue may be maintained.

By means of the presence of Natrum Phos. in the system, lactic acid is decomposed into carbonic acid and water. This salt has the power of holding carbonic acid in combination, fixing it, and does this

4

in the proportion of two parts of carbonic acid to one of phosphoric acid which it contains. This combination is carried to the lungs, and there, by the action of oxygen from the inhaled air, the carbonic acid is set free from its loose union with Natrum Phos.—is exhaled and exchanged for oxygen.

Thus, of the sodiums, it will be noted that Natrum Phos. creates, Natrum Mur. distributes, and Natrum Sulph. eliminates water.

The final products of the oxidation of the organic substances are urea, carbonic acid and water. These, together with salts set free, leave the tissues and thereby give place to less fully oxidized organic bodies, which in turn undergo finally the same metamorphosis.

The products of this retrograde tissue change are conveyed through the lymphatics, the connective tissue and the veins to the gall-bladder, lungs, kidneys, bladder and skin, and are thereby removed from the organism with the excretions, such as the urine, perspiration, faeces, etc.

The human organism can be said to be in a state of health when the cells of the body are acting normally. Cell action is dependent for its nourishment on the daily consumption of suitable food and drink. When this is properly digested and assimilated the blood is compensated for the losses sustained by supplying nutritive material to the tissues, to carry on the activities of life. When this nutritive material is supplied in requisite quantities and there exists no disturbance in the rhythmical motion of the molecules then the building of new cells and the destruction of the old cells, together with the elimination of waste products continues normally and the condition called health proceeds uninterruptedly.

But if from any known or unknown cause there takes place a pathogenic irritation of the cell its activity is at first increased, due to its effort to resist this irritation. However, through this effort to repel the irritation it may lose a portion of its mineral constituents and then undergo a pathogenic change.

Describing the change of cell due to pathogenic irritation, Schuessler says: "Suppose the functional material lost in the contest with the pathogenic irritation to be, e. g., *Potassium Chloride*, then it has also lost a corresponding quantity of fibrin, for *Potassium Chloride* and fibrin have a physiologico-chemical relationship. If the cell in its contest with the pathogenic irritation has lost *Calcium Phosphate*, it has also lost a corresponding quantity of albumen, because *Calcium Phosphate* has a similar relation to albumen as *Potassium Chloride* has to fibrin. An exudation of fibrin, therefore, presupposes a deficiency of *Potassium Chloride* and an exudation of albumen presupposes a deficiency of *Calcium Phosphate* in the cells immediately contiguous to the exudation referred to. A loss in the other cell-minerals may be similarly deduced.

The cells which have undergone pathogenic changes, *i. e.*, the cells in which there is a deficiency in one of their mineral constituents, need a compensation by means of a homogeneous mineral substance. Such a compensation may be made spontaneously, *i. e.*, through the curative effort of Nature whereby the requisite substances enter the cells from their interstices. But if the spontaneous cure is delayed, therapeutic aid becomes necessary."

This therapeutic aid, mentioned by Schuessler, consists in furnishing to the diseased cells a minimal dose of that inorganic substance, or cell-salt, whose molecular motion is disturbed, which disturbance caused the diseased action. To do this successfully it is necessary to know what salts are needed for the upbuilding of the different tissues and for their normal action. This knowledge is derived from physiological chemistry, and hence this treatment of disease by supplying the needed cell-salts is called the biochemical treatment.

In the following pages are given, under the different names of diseases, the respective cell-salts that will prove curative, based upon the kind of cells affected by the different diseases. Thus, in catarrhal conditions, for instance, the remedial cell-salts will be the same, whether the catarrh shows itself in the throat, nose, or other organs, since it is the mucous membrane that is involved, and the mucous cells, therefore, call for a cell-salt that is lacking.

In order that these lacking salts may be assimilated by the cells, however, it is essential that they should be prepared in strict accordance with the biochemical method.

— ... —

PREPARATION AND DOSE OF THE TISSUE SALTS

The best form of the tissue salts (sometimes refered to as the cell salts) is the celloid or tablet form. This is obtained in the following manner: The original natural salt is triturated according to the biochemical method with sugar of milk, one part of the salt to nine of sugar of milk, for at least one hour; this gives the first decimal trituration of "1 X". The particles of this are still too large to readily assimilated by the cells of

7

the body. Experience has shown that for general use the third (3 X) or sixth (6 X) trituration are the more desirable. (One one thousandth or one one millionth.) From these triturations the celloids or triturate tablets are made. With the celloids the dosage can be more easily controlled than with the powder.

— · · · —

Celloids of the tissue remedies should be prescribed in doses of 3 to 5 tablets, taken dry on the tongue, or if preferred, dissolved in water, every half hour or hour, according to the severity of the case, and every three hours after amelioration. When two or more salts are needed they should be taken in alternation. In chronic cases two doses a day should be given, in the morning on rising, and at night before retiring. If the pains or symptoms are very severe the quickest results will be obtained by dissolving the medicine in hot water and so administering it until relief is obtained.

Nature works everywhere with an immense number of infinitely small atoms which can only be perceived by our dull organs or senses when presented to them in finite masses. The smallest image our eye can see is produced by millions of waves of light. A granule of salt which we can scarcely taste contains millions and millions of groups of atoms which no human eye will ever discern.

One quart of milk is found by analysis to contain about the six-millionth of a grain of iron; a child fed on milk receives each time one milligramme of iron in a half-pint of milk, which is only the fourth part of the above minute fraction of one part of a grain of iron.

8

Four milligrammes represents the whole quantity of iron in the milk supplied per day for its nourishment and growth, and this is sufficient to feed all the cells that are known to contain iron, and consequently require iron. This being the fact, how small will be the quantity required to equalize the balance of iron molecules in only a limited portion of a group of cells, where for instance, a molecular disturbance has taken place, and iron has to be supplied medicinally. But if milk contains the whole of the twelve inorganic cell-salts, how small must the quantity be when subdivided so that each drop has its own particle of each of the twelve constituents.

The proportion of fluorine in the human organism is still less than that of iron. From analytical facts it may be estimated that the fluorine in milk is only present in decimilligrammes. One milligramme of Calcarea Fluor. per dose for a remedy would be quite large. A dose of any remedy used for therapeutic purposes should be rather too small than too large; for if too small, a repetition of the dose will bring about the desired effect, while too large a dose may miss its object altogether. Large doses of iron have a bad effect on the stomach, leaving the complaint unaffected. At the temperature of the body, hydrochloric acid, diluted with one-thousandth part of water, readily dissolves the fibrin of meat and the gluten of cereals, and this solvent power is decreased, not increased, when the acid solution is made stronger. (Professor Liebig's Chemical Letters.)

Spectrum analysis has opened a new field of truth, showing matter to be capable of endless subdivision.

A disturbance in the molecular movements of any of the inorganic salts of a tissue produces an altered or abnormal condition, which is termed disease. Professor Virchow, the greatest authority of the day on cellular diseases and cancer cells, clearly states that the definition of all disease resolves itself into this: "An altered or changed state of cell." For the healing or cure of such Dr. Schuessler supplies the smallest dose of the identical inorganic substance, because the molecules of that substance, administered as medicine, fill up the gap in the chain of molecules of that particular cell or tissue-salt. Chemical affinity plays here a particular part, each salt, by virtue of that law existing between organic and inorganic substances, finding its way into its particular tissue where it is wanted. Under this law nature cures; hence it becomes necessary to administer these salts to the minute cells medicinally in minute quantities. Thus refined, they can be taken up by the cells so changed that they are no longer able to absorb the ordinary molecules of salts out of the plasma. Hence it follows that the ordinary preparations of cell-salts given as medicines are too bulky, and Dr. Schuessler has formulated a saccharated trituration of the twelve constituents of the body in such form that they can pass by their special passages in the capillaries and are readily assimilated by the cells of the blood and tissues. One illustration explains this: One red blood corpuscle does not exceed the one-hundred-and-twenty-millionth of a cubic inch. There are over three million such cells in one droplet of blood, and these cells carry the iron in the blood. How necessary, then, to administer the cell-salt iron (Ferrum phos., which is the remedy for all inflammations of lungs, pleura, throat, eyes, ears, etc., included) to diseased cells in the most minute molecular form. Each one of the twelve inorganic substances (in chemistry called salts) of which the human body is built up, has its own sphere of function and curative action, by reason of the part it occupies in the cells, and the part these have to perform to maintain and restore health.

THE TWELVE TISSUE-SALTS

THEIR PLACE AND FUNCTION
IN THE HUMAN ECONOMY

1. CALC. FLUOR.
   (*Calcium Fluoride*)

2. CALC. PHOS.
   (*Calcium Phosphate*)

3. CALC. SULPH.
   (*Calcium Sulphate*)

4. FERR. PHOS.
   (*Phosphate of Iron*)

5. KALI MUR.
   (*Potassium Chloride*)

6. KALI PHOS.
   (*Potassium Phosphate*)

7. KALI SULPH.
   (*Potassium Sulphate*)

8. MAG. PHOS.
   (*Magnesium Phosphate*)

9. NAT. MUR.
   (*Sodium Chloride*)

10. NAT. PHOS.
    (*Sodium Phosphate*)

11. NAT. SULPH.
    (*Sodium Sulphate*)

12. SILICA
    (*Silicic Oxide*)

IN ORDER to obtain the most satisfactory results from Dr. Schuessler's Biochemic System of Medicine, one should first of all become acquainted with the tissue-salts individually. A knowledge of the properties and the field of action of each of these remedies will be found invaluable when the symptoms of any given case are be-ing considered. In the following brief review, an attempt has been made to bring out the dominant characteristics

11

of each of the tissue-salts and thus to make quicker and easier the right choice of remedy.

CALC. FLUOR. (Calcium Fluoride)

Calc. Fluor. gives to the tissues the quality of *elasticity*. It combines with the organic substance, albumin, to form organic elastic tissue and is found in the walls of the blood vessels, in muscular tissue, in connective tissue, in the surface of bones and in the enamel of teeth. A deficiency of *Calc. Fluor.* results in a loss of elasticity and consequent relaxed condition. Its main function is the preservation of the contractile power of elastic tissue.

Whenever symptoms are traceable to a relaxed condition this tissue-salt is indicated, e.g. a relaxed condition of veins and arteries, piles, sluggish circulation, a tendency to cracks in the skin, notably in the palms of the hands and between the toes. *Calc. Fluor.* is also useful in the treatment of diseases affecting the surface of the bones and joints and when the teeth become loose in their sockets and decay rapidly. The elasticity of muscular tissue and supporting membranes becomes impaired when this tissue-salt is deficient, resulting in muscular weakness, bearing-down pains, etc. The symptoms are generally worse in humid conditions and are relieved by massage and warmth.

CALC. PHOS. (Calcium Phosphate)

Calc. Phos. is the tissue-salt concerned with *nutrition*. It combines with albumin and is indicated when there are albuminous discharges. Without *Calc. Phos.* there could be no blood coagulation. It will assist the action of a more directly indicated tissue-salt and thus produce more rapid results. It promotes healthy cellular activity and restores tone to weakened organs and tissues. This tissue-salt is concerned with the formation of bone and teeth and thus becomes an important remedy for children. It aids growth and normal development and should be given in cases of backwardness: more particularly where there is bone weakness or recurring tooth troubles.

Calc. Phos. is the biochemic remedy for rickets. It is a constituent of saliva and gastric juice. It assists digestion and assimilation and favours the building up of a sturdy, robust constitution. This is the remedy for the period of convalescence, its restorative power will speed recovery and replenish the body's reserves of strength. *Calc. Phos.* is the tissue-salt for blood poverty and conditions associated with imperfect circulation. In the anæmic states often seen in young girls, this remedy should be given. *Calc. Phos.* pains can be severe and "fixing" and they tend to be worse at night. There may be a creeping sensation of the skin, also numbness and coldness of the limbs. *Calc. Phos.* has always been prized as a restorative.

CALC. SULPH. (Calcium Sulphate)

Calc. Sulph. is a *blood purifier* and *healer*. It is found in the liver where it helps in the removal of waste products from the blood stream and it has a cleansing and purifying influence throughout the system. *Calc. Sulph.* cleans out the accumulation of non-functional, organic matter in the tissues and causes infiltrated parts to discharge their contents readily, throwing-off decaying organic matter, so that it may not lie dormant or slowly decay and thus injure the surrounding tissues.

Calc. Sulph. is indicated in conditions arising from impurities in the blood stream. It supplements the action of *Kali Mur.* in the treatment of catarrh, acne, etc., and it should always be given when "pimples" occur in adolescence. It checks the weakening drain of suppuration too long continued, e.g. abscesses and wounds which will not heal readily and tend to become septic. If taken

13

in the early stage, it will prevent a sore throat from developing and in the same way, it will often cut short a threatening cold. The symptoms are generally worse after getting wet and are better in a warm, dry atmosphere.

FERR. PHOS. (Iron Phosphate)

Ferr. Phos. is the pre-eminent Biochemic First-Aid. It is the *oxygen-carrier*. It enters into the composition of hæmoglobin, the red colouring matter of the blood. It takes up oxygen from the air inhaled by the lungs and carries it in the blood stream to all parts of the body thus furnishing the vital force that sustains life. It gives strength and toughness to the circular walls of the blood vessels, especially the arteries. Freely circulating, oxygen-rich blood is essential to health and life and for this reason *Ferr. Phos.* should always be considered, as a supplementary remedy, no matter what other treatment may be indicated by the symptoms.

Congestion, inflammatory pain, high temperature, quickened pulse, all call for more oxygen, and it is *Ferr. Phos.* that is the medium through which oxygen is taken up by the blood stream and carried to the affected area. This tissue-salt can be given with advantage in the early stage of most acute disorders, and it should be administered at frequent intervals until the inflammatory symptoms subside. It is also indicated where there is a lack of red blood corpuscles, as in anæmia, and as a first-aid remedy for hæmorrhages. It would be difficult to find a case of illness where *Ferr. Phos.* could not be used to advantage, irrespective of any other treatment that may be given. It is an excellent remedy for ailments associated with advancing years and it is one of the most frequently needed remedies in the treatment of children's ailments. Bleeding from wounds, cuts and abrasions, can be controlled with a little powdered *Ferr. Phos.*, applied direct to the injured parts. A few tablets

14

may be crushed for this purpose or the tablets may be dissolved and used as a lotion (see directions, external applications). *Ferr. Phos.* should also be thought of as a first-aid in cases of muscular strains, sprains, etc.

NOTE: *Ferr. Phos.* is in no sense an iron tonic. Its action is entirely nutritional and without any side-effects.

KALI MUR. (Potassium Chloride)

Kali Mur. is the remedy for sluggish conditions. It combines with the organic substance, fibrin. Thus a deficiency of this tissue-salt causes fibrin to become non-functional, and to be thrown off in the form of thick, white discharges, giving rise to catarrhs and similar symptoms affecting the skin and mucous membranes. Its action is complementary to that of *Calc. Sulph.*, as both remedies are concerned with cleansing and purifying the blood. In conditions calling for *Kali Mur.* the blood tends to thicken and to form clots. In alternation with *Ferr. Phos.* it is frequently needed for the treatment of children's ailments.

Kali Mur. is the remedy for thick, white fibrinous discharges. Other prominent symptoms are a white-coated tongue and light-coloured stools (lack of bile). Torpidity of the liver is another indication. In alternation with *Ferr. Phos.* it is frequently needed in the treatment of inflammatory diseases, particularly those affecting respiration—coughs, colds, sore throats, tonsillitis, bronchitis, etc.; also for children's ailments such as measles and chicken pox and where there are soft swellings, e.g. mumps, croup. *Kali Mur.* is concerned with the production of saliva and is therefore important in the early stages of digestion. The symptoms may be worse after eating fatty or rich foods and there may be lack of appetite. With *Nat. Mur.* it is utilised in the production of hydrochloric acid, and is thus an essential link in the process of digestion. This tissue-salt is useful as a first-aid for the treatment of burns.

KALI PHOS. (Potassium Phosphate)

Kali Phos. is a *nerve nutrient*. It is the remedy for ailments of a truly nervous character. School children often need this tissue-salt; it helps to maintain a happy, contented disposition and sharpens the mental faculties. Early symptoms may be very slight, scarcely noticeable in fact, except to a mother's watchful eye. There may be fretfulness, ill-humour, bashfulness, timidity, laziness and similar indications; indeed, any display of what is sometimes described as "tantrums" may be regarded as a *Kali Phos.* symptom.

Kali Phos. is the remedy for nervous headaches, nervous dyspepsia, sleeplessness, depression, languid weariness, lowered vitality, grumpiness and many other conditions which may be summed up in the modern colloquial phrase, "lack of pep". But do not regard *Kali Phos.* as merely a pick-me-up, this tissue-salt is an important constituent of nervous tissue and consequently has a wide and powerful influence on the bodily functions. It covers those ailments comprehended by the term "nerves". *Kali Phos.* is also indicated in the treatment of irritating skin ailments, such as shingles, to correct the underlying nervous condition. It is helpful for breathing in nervous asthma. The symptoms are usually worse from mental and physical exertion and from cold. They are ameliorated by rest, warmth and sometimes by eating.

KALI SULPH. (Potassium Sulphate)

Kali Sulph. works in conjunction with *Ferr. Phos.* as an oxygen-carrier. It assists in the exchange of oxygen from the blood stream to the tissue-cells, thereby completing the respiratory process initiated by *Ferr. Phos.* *Internal* breathing of the tissues depends upon *Kali Sulph.*; *external* breathing is the function of *Ferr. Phos.*, if we designate the exchanges of gases in the lung in this way. *Kali Sulph.* has a beneficial effect on respiration and is indicated in those cases where there is a feeling of "stuffiness" or desire for cool air. It is also the anti-friction salt ensuring the smooth working of all parts, thus acting in the manner of a lubricant.

Kali Sulph. is indicated where there is a sticky, yellowish discharge from the skin or mucous membrane, as in certain forms of catarrh. Eruptions on the skin and scalp, with scaling, call for this remedy and it helps to maintain the hair in a healthy state. Other symptoms include chilliness and shifting, fleeting pains. It is useful in the treatment of intestinal disorders, stomach catarrh, and in inflammatory conditions to promote perspiration. The symptoms are generally worse in the evening, or in a closed, stuffy atmosphere, and are better in the fresh air.

MAG. PHOS. (Magnesium Phosphate)

Mag. Phos. is known as the *anti-spasmodic* tissue-salt. Its main function is in connection with the nervous system where it supplements the action of *Kali Phos.* When a deficiency of *Mag. Phos.* occurs, the white nerve fibres contract, causing spasms and cramps. This tissue-salt is of importance to muscular tissue ensuring rhythmic and coherent movement. *Mag. Phos.* is quick to relieve pain, especially cramping, shooting, darting or spasmodic pains.

Mag. Phos. is indicated for nerve pains, such as neuralgia, neuritis, sciatica, and headaches accompanied by shooting, darting stabs of pain. It relieves muscular twitching, cramps, hiccups, convulsive fits of coughing and those sudden, sharp twinges of pain that are so distressing. It also relieves menstrual pains. Stomach cramps and flatulence respond to treatment with this tissue-salt. These symptoms may be aggravated by cold and by touching and are relieved by the application of heat, by pressure and by bending double. The doses may be taken at frequent intervals until relief is obtained.

Because magnesium has been assumed to be plentiful

17

in the diet, some authorities have considered a deficiency unlikely. But this is not necessarily so, as recent research has proved that some diets provide insufficient magnesium for the body's needs.

NOTE:—*Mag. Phos. will often act more rapidly when the tablets are taken with a sip of hot water.*

NAT. MUR. (Sodium Chloride)

Nat. Mur. is the *water-distributing* tissue-salt. It enters into the composition of every fluid and solid of the body. Because of its powerful affinity for water, it controls the ebb and flow of the bodily fluids; its prime function being to maintain a proper degree of moisture throughout the system. Without this tissue-salt, cell division and normal growth could not proceed. It is closely associated with nutrition, with glandular activity and with the internal secretions which play such an important part in the physiological process. Excessive moisture or excessive dryness in any part of the system is a clear indication of a *Nat. Mur.* deficiency. The resulting symptoms are many and varied but always, underlying them, will be found this predominant condition of too much or too little water. Here are some typical symptoms:

Low spirits, with a feeling of hopelessness; headaches with constipation; blood thin and watery with pallor of the skin, which sometimes has a greasy appearance; difficult stools, with rawness and soreness of the anus; colds with discharge of watery mucus and sneezing; dry, painful nose and throat symptoms; heartburn (waterbrash) due to gastric fermentation with slow digestion, the food remains too long in the stomach; great thirst; tooth-ache and facial neuralgia with flow of tears and saliva; eyes weak, the wind causes them to water; hay fever, drowsiness with muscular weakness; chafing of the skin; hang-nails; unrefreshing sleep—tired in the morning; after-effects of alcoholic stimulants; loss

18

of taste and smell; craving for salt and salty foods; stings and bites of insects—apply locally as soon as possible.

An important function of *Nat. Mur.* is the production of hydrochloric acid. Too little acid means slow digestion, especially of calcium rich foods.

Remember, approximately two thirds of your body is composed of *water*; hence the vital role played by *Nat. Mur.*, the water distributor, in all the life processes.

NAT. PHOS. (Sodium Phosphate)

Nat. Phos. is an *acid neutraliser*. It is the principal remedy for the wide group of ailments arising from an acid condition of the blood. This tissue-salt is also of importance for the proper functioning of the digestive organs. The assimilation of fats and other nutrients is dependent on the action of this remedy. A deficiency of *Nat. Phos.* allows uric acid to form salts which become deposited around the joints and tissues giving rise to stiffness and swelling, and other painful rheumatic symptoms.

Nat. Phos. is indicated whenever symptoms of acidity are present, such as acid dyspepsia, pain after eating and similar digestive disorders. Other indications are highly coloured urine, golden-yellow or creamy coating at the root of the tongue (the whole tongue may sometimes present the appearance of a piece of washleather), worms, nervous irritability. Sleeplessness caused by indigestion can sometimes be remedied with a dose of *Nat. Phos.* kept handy by the bedside. This remedy is of importance in the treatment of rheumatism, lumbago, fibrositis and associated ailments. An acid state of the blood occurs when there is a deficiency of the soothing, acid-neutralising tissue-salt, *Nat. Phos.*

NAT. SULPH. (Sodium Sulphate)

Nat. Sulph. regulates the *density* of the intercellular fluids (fluids which bathe the tissue-cells) by eliminating

19

excess water. This tissue-salt largely controls the healthy functioning of the liver; it ensures an adequate supply of free-flowing, healthy bile, so necessary for the later stages of digestion. The removal of poison-charged fluids, which are the normal result of the chemical exchanges constantly taking place in the tissue-cells, is brought about by the action of *Nat. Sulph.* If conditions arise which allow these waste fluids to accumulate in the blood and tissues, auto-intoxication (self-poisoning) is the result. *Nat. Sulph.* ensures the disposal of these poison-charged fluids and its importance in the treatment of rheumatic ailments is therefore self-evident.

Nat. Sulph. is indicated in the treatment of ailments affecting the liver, e.g. biliousness. Sandy deposits in the urine, watery infiltrations, a brownish-green coating of the tongue and a bitter taste in the mouth are some of the symptoms. It is the principal remedy in the treatment of influenza. Humid asthma, malaria and other conditions associated with humidity need this remedy. A few doses of *Nat. Sulph.* will help to dispel that languid feeling so often experienced during a spell of humid, oppressive weather.

SILICA (Silicic Oxide)

Silica is a *cleanser* and *eliminator*. It is a deep-acting remedy which helps the body to throw off non-functional organic matter that may have arrived at a given point during Nature's effort to eliminate it from the system. It can often initiate the healing process by promoting suppuration and breaking up pathological accumulations, e.g. abscesses. *Silica* is a constituent of the hair, skin, nails and surfaces of the bones. It also acts in the manner of an insulator for the nerves. In cases of checked perspiration, *Silica* restores the activity of the skin, thereby aiding this important cleansing process. It is the biochemic remedy for offensive perspiration of the feet and arm-pits.

SELECTING THE REMEDY

SYMPTOMS ARE the body's warning signals. They are the pointers which, if carefully noted, will indicate the remedy required. Symptoms occur in a variety of forms, such as pain, inflammation, swelling, exudation and so on, and their nature and location determines the particular tissue-salt called for. For example, in the case of a chill the symptoms may be a watery discharge from the nose, loss of sense of smell or taste, dryness of the bowel, or excessive thirst, all of which may arise from a common cause—a deficiency of the moisture regulating tissue-salt, *Nat. Mur.* and that would be the remedy to be given where these symptoms are predominant.

Having carefully noted all the symptoms, the next step in selecting the remedy is to compare them with the guiding symptoms and characteristic indications of each of the tissue remedies. It will usually be found that one remedy corresponds most closely to the case for which treatment is required. That tissue-salt is known as the principal remedy. Other remedies may also be indicated and these are known as supplementary remedies. Each remedy has its own characteristic symptom, e.g. *Ferr. Phos.* inflammation; *Mag. Phos.* spasmodic pain and cramp; *Nat. Phos.* acidity; *Calc. Phos.* ill-nourished states, and so forth. These key symptoms are readily recognised and give a sure lead to the treatment called for in any given case.

The key to success in Biochemic therapy lies in the accurate linking of the symptom with its appropriate tissue-salt. Facility in achieving this can be acquired by any intelligent person giving the necessary time and thought to become familiar with the sphere of action of the individual tissue-salts.

CHANGING THE REMEDY

In the treatment of some ailments, particularly those of an acute kind, differing symptoms may make their appearance during the course of treatment, and in such cases the remedies should be changed or supplemented in accordance with the variation of the symptoms at each successive stage.

CLEANSING THE SYSTEM

It has already been explained that the inorganic constituents of the human system—the twelve tissue-salts—are the tissue builders which combine with and convert organic matter into living tissue. Thus any tissue-salt deficiency or imbalance renders its organic counterpart non-functional. The system seeks to eliminate this useless material and symptoms such as catarrh and eczema are the visible results of this natural, cleansing process. Clearly, such symptoms should never be suppressed, otherwise this waste matter, by clogging the system, would soon bring the life processes to a halt. The tissue-salts to be thought of in such cases are those principally concerned with catarrhal conditions, notably, *Calc. Sulph.* to purify the blood, *Kali Mur.* to correct sluggish, congested conditions, and *Nat. Sulph.* the liver salt.

CORRECTING PAST ERRORS

Numbers of people are unknowingly suffering from the cumulative effects of suppressive treatments to which they have been subjected in the past—medicinal drug poisoning—and, in such cases, response to biochemic treatment may be slow. If little or no benefit is apparent after a reasonable period of treatment with the remedy indicated by the symptoms, the system can be rendered more responsive by a short anti-toxic course consisting of the two remedies, *Nat. Mur. and Kali Sulph.* in alternation; three doses of each remedy taken daily. After this corrective treatment other tissue-salts will usually prove fully effective in their own sphere of action.

TO the newcomer to Biochemistry the list of ailments and suggested remedies given in the following section will prove helpful. But, as experience in the use of the tissue-salts is acquired, the aim should be to determine the most prominent symptom in any given case and to treat that symptom first. The name of an ailment can mean little or nothing, but symptoms, viewed in the light of the teachings of biochemistry, are a guide to the nature of the body's deficiency and to the corrective treatment required. It should be borne in mind that these suggestions are intended as first-aids for the treatment of minor ailments. It is always advisable to consult a doctor, preferably a homœopathic physician, whenever the symptoms persist, or if they are in any way unusual at the beginning or during the course of an illness. As a means of providing first-aid, biochemic treatment is invaluable. The tissue-salts are perfectly safe at all times and they cannot conflict with other treatments.

BIOCHEMIC FIRST-AID

COMMON MINOR AILMENTS
(arranged alphabetically)

ABSCESSES

An acute abscess is one which develops rapidly, beginning as a sore spot in some part of the body, becoming hard, inflamed, painful and filled with pus. When pus formation has occurred it is an indication that the white blood corpuscles have successfully overcome the invading micro-organisms. There may be some fever. Hot fomentations will help to relieve the pain, which subsides when the abscess bursts. Boils resemble abscesses, but usually come in crops. Meticulous cleanliness of the affected area is important in preventing the spread of infection. The action of *Silica* helps the abscess to ripen. A lowered state of health is sometimes a predisposing factor and in such cases a course of *Calc. Phos.* is indicated.

BIOCHEMIC TREATMENT

Ferr. Phos. The first remedy for the pain, heat and threatened suppuration.

Kali Mur. When there is swelling but no pus formation. It should be given in the early stages in alternation with *Ferr. Phos.*

Silica. Helps the abscess to ripen and discharge its contents readily. Should be given when suppuration appears.

Calc. Sulph. In the final stages for cleansing and healing.

26

Kali Phos. is antiseptic in action, and is therefore useful as an intercurrent remedy.

ACIDITY

Acidity is a somewhat loose term indicating that the blood, or one or more of the secretions, is less alkaline than it should be. This excess of acid gives rise to many distressing symptoms. There may be gastric disturbance, irritation of the skin and mucous membranes, impoverishment of the blood, palpitation of the heart, twinges of rheumatism, headache on the top of the head with a sense of fullness, a persistent feeling of tiredness and other symptoms of disturbed metabolism. Whenever signs of acidity make their appearance the principal remedy, *Nat. Phos.*, should be given—irrespective of any other treatment which may be indicated—as this acid state, if allowed to persist, will hamper the action of other tissue-salts.

BIOCHEMIC TREATMENT
Nat. Phos. The principal remedy whenever symptoms of acidity are present.

Nat. Sulph. This tissue-salt is one of the alkaline sulphates and it may be used to supplement the action of *Nat. Phos.*

Silica. Dyspepsia with eructations, heartburn, chilliness. In alternation with *Nat. Phos.*

Mag. Phos. Burning, tasteless eructations, relieved by drinking hot water. Flatulence with distension of the stomach, belching of gas and full sensation in the abdomen.

ANÆMIA

Anæmia is an impoverished condition of the blood. The blood-cells may be too few in number (*Calc. Phos.*)

27

or the oxygen-carrier hæmoglobin may be deficient (*Ferr. Phos.*). The simpler forms of anæmia may arise from a variety of causes, bleeding, after childbirth, a deficient diet and so on. Foods rich in vitamins such as red meats, liver, milk, eggs and green vegetables should be included in the diet. Fresh air and sunshine are also valuable aids in restoring the quality of the blood.

There is a form of anæmia appearing in young girls at the age of puberty and this may be recognised by the resulting pallor of the skin. The girl loses weight, is excessively tired, has difficulty in keeping warm, has no appetite, is usually constipated and the appearance of the menstrual period is delayed or becomes irregular. Medical advice should be sought.

BIOCHEMIC TREATMENT

Calc. Phos. The principal remedy to provide new blood-cells. Especially useful for anæmic children and during convalescence.

Ferr. Phos. Helps in the formation of red blood by bringing oxygen to the new blood-cells.

Kali Mur. If skin eruption exists, or the stools are light coloured.

Nat. Mur. When the blood is thin and watery with depression of spirits and prostration. *Nat. Mur.* is an important remedy in the treatment of anæmia, as cell reproduction is arrested if this remedy is deficient.

Nat. Phos. Useful as an intercurrent remedy and when an acid condition prevails.

ASTHMA

Asthma is a respiratory disorder characterised by paroxysms of difficult breathing, usually followed by periods of relief, with recurrence of the attacks at fairly regular intervals. There are various kinds of asthma and sufferers should seek medical advice. An attack may be brought on by contact with substances to which the person is sensitive and emotional or dietary factors should not be overlooked. A sudden change of temperature can bring on an attack. Asthma is more common in men than in women and the first attack usually occurs in childhood. *Some alleviation of the symptoms may be obtained with the use of the appropriate tissue-salts.*

Biochemic Treatment

Kali Phos. Nervous asthma, hay asthma. The chief remedy for the breathing and depressed nervous state.

Mag. Phos. Spasmodic nervous asthma. In alternation with *Kali Phos.*

Kali Mur. With gastric derangement, tongue coated white and mucus white.

Nat. Mur. Profuse, frothy mucus and tears streaming when coughing.

Calc. Phos. Bronchial asthma; clear, tough, gluey expectoration.

Kali Sulph. Bronchial asthma with yellow expectoration. Worse in the evenings or in a hot, stuffy atmosphere.

Nat. Sulph. Asthma due to humid conditions with greenish, copious expectoration.

BACKACHE

Backache is a symptom of many ailments. It may be due to local causes, such as lumbago, rheumatism, strains, etc. On the other hand the trouble may be more deep-seated and medical advice should be sought if the trouble persists.

Biochemic Treatment

Ferr. Phos. Acute inflammatory pains in the loins.

Kali Mur. Useful intercurrently with *Ferr. Phos.* and for pain during the menstrual periods.

Calc. Fluor. Bearing-down pains in the lower part of the back with tired feeling.

Mag. Phos. Boring, darting, neuralgic pains in any part of the back.

BED WETTING

Bed wetting (enuresis) is a fairly common habit with some young children. It may be due to nervousness (*Kali Phos.*) indigestion (*Nat. Phos.*), too much liquid before going to bed or just a habit. It is best to avoid giving liquid for at least an hour before bedtime. The urine should be voided just before going to bed as some youngsters are scared of getting up in the dark. A night-light may be helpful.

BIOCHEMIC TREATMENT

Ferr. Phos. If inflammation is present and for muscular weakness.

Kali Phos. For nervous, highly strung children.

Nat. Mur. In alternation with *Ferr. Phos.* or *Kali Phos.* as indicated.

Nat. Phos. When accompanied by symptoms of acidity.

BILIOUSNESS

Bile is a bitter, yellowish fluid secreted by the liver and stored in the gall bladder. It is discharged through the bile duct into the intestine where it assists in the process of digestion and assimilation. About a pint or more is secreted daily, but much of this is reabsorbed into the bloodstream and circulates back to the liver, to be again excreted, and so on. Biliousness is a rather vague term applied sometimes to migraine or to the sick headache and vomiting which occur in some forms of gastric catarrh or following indiscretions of diet. The liver salt, *Nat. Sulph.*, is the principal remedy indicated for disorders in the secretion and flow of bile. Attention should also be given to the diet, and foods and drinks known to disagree should be avoided. Lemon juice may help dispel the nausea.

BIOCHEMIC TREATMENT
Nat. Sulph. The principal remedy for the nausea, vomiting and giddiness.
Kali Mur. When associated with digestive disorders with white-coated tongue or light-coloured stools.
Ferr. Phos. For sick headache and soreness of scalp in alternation with the principal remedy.
See also Combination S.

BOILS (See Abscesses)

BRONCHITIS

Bronchitis is an inflammatory condition of the mucous membranes of the bronchial tubes. It may follow a cold or catarrh. The usual symptoms are feverishness with a harsh, dry cough and wheezing respiration. The painful chest symptoms become less distressing when expectoration begins. Rest in bed and a light diet, with warm drinks, are beneficial. If the inflammation spreads into the smaller bronchial tubes the symptoms become intensified with respiration rapid and difficult. Special care should be taken in the case of the very young and the aged. In the earlier stages, relief can usually be obtained by the inhalation of vapour with a little friars balsam and menthol added to the steaming water. Medical advice should be obtained.

BIOCHEMIC TREATMENT
Ferr. Phos. The first remedy for the inflammation and temperature.

31

Kali Sulph. In alternation with *Ferr. Phos.* to promote perspiration and control fever. It should be continued if expectoration is yellow and slimy, and for evening aggravation of symptoms.

Kali Mur. For the second stage with thick, white phlegm, feeling of stuffiness, and whitish-grey tongue.

Nat. Mur. Watery, frothy expectation with loss of taste and smell.

Calc. Sulph. A useful remedy, in alternation with *Calc. Phos.*, to speed recovery during convalescence.

CATARRH

An excessive secretion from the mucous membranes, particularly those of the air passages. It usually begins as a nasal catarrh (coryza) with a feeling of stuffiness and sneezing, accompanied by a profuse, watery discharge from the eyes and nose. Sometimes there is a loss of sense of smell and taste. There may also be soreness of the throat with bronchial irritation and symptoms similar to those of the common cold. The nature of the discharge is a guide to the tissue-salts required. The name catarrh is also applied to describe similar conditions affecting the stomach and bowels.

BIOCHEMIC TREATMENT

Ferr. Phos. In the first stage for the fever and congestion.

Kali Mur. For the second stage when there is thick, white phlegm and a feeling of stuffiness. For chronic catarrh causing difficulty in breathing.

Kali Sulph. An important tissue-salt in the treatment of affections of the mucous membranes with sticky, yellowish secretions. Catarrh of the stomach. This remedy works well with *Ferr. Phos.*

Nat. Mur. Catarrh and colds with watery, transparent, frothy discharges. Dryness of the nose with loss of taste and smell.

CHICKEN POX

Chicken pox is an acute contagious disease of children characterised by feverishness and an eruption on the skin. It occurs in epidemics and the incubation period can vary from eleven to twenty-one days after exposure to infection. The first symptom is feverishness with aching of the back and legs. Within twenty-four hours there is an eruption of red pimples, which later become filled with a clear fluid. These vesicles eventually form small crusts which scale off in little more than a week. The child must be isolated from other children for fourteen days from the appearance of the rash. Scratching should be discouraged or the marks on the skin may remain. In some cases shingles in adults is thought to be due to infection from chicken pox. Medical advice should be sought.

BIOCHEMIC TREATMENT
Ferr. Phos. This remedy should be given every hour from the onset of the fever until the feverishness subsides.

Kali Mur. In alternation with *Ferr. Phos.* during the eruptive period.

Kali Sulph. Important as a supplementary remedy for the scaling of the skin.

Calc. Phos. To be given during convalescence.

CHILBLAINS

An inflamed condition of the skin with swelling of the subcutaneous tissues, usually affecting the hands or

33

feet. They occur in persons having a defective circulation and sometimes there is a nutritional deficiency (*Calc. Phos.*). A nourishing diet and warm clothing are obvious preventive measures, and regular exercise and massage, by helping to improve the circulation, remove the predisposing conditions.

BIOCHEMIC TREATMENT

Calc. Phos.   This is the principal remedy.

Kali Mur.   Useful where there is much swelling.

Ferr. Phos.   In alternation with *Kali Mur.* for the pain and inflammation.

Kali Phos.   May be given intercurrently to counter the effects of the irritation.

Calc. Fluor.   A useful remedy when there are cracks in the skin.

Kali Sulph.   Broken chilblains exuding thin, yellow fluid.

COLDS

Although, generally speaking, colds are regarded as trivial everyday ailments, they can lead to serious trouble. It is therefore a wise precaution never to neglect the common cold. If you can go to bed for a day or two, do so. You will benefit from the rest and your system will have a better chance of dealing with this temporary indisposition. Colds are, in effect, a cleansing process, indicating that the body is ridding itself of non-functional organic matter. Do not attempt to suppress a cold. Assist the process of elimination by giving copious drinks, lemon, barley water, etc., and keeping the bowels active, thereby giving the digestive system a well-earned respite. These common-sense measures should serve to cut short the course of the cold and other people will not be exposed to infection. If the symptoms persist, medical advice should be sought, as more serious troubles sometimes start with the same symptoms as a cold.

People who are susceptible to colds will find a course of *Ferr. Phos.*, *Kali Mur.* and *Nat. Mur.* is helpful.

BIOCHEMIC TREATMENT

Ferr. Phos. is needed for the feverishness, stuffiness and sneezing that herald the onset of a cold. That bout of sneezing is a first sign that a cold is threatening. It is a clear and unmistakable call for *Ferr. Phos.* Recognise that fact and act accordingly and you will be spared much discomfort and inconvenience.

Kali Mur. For the second stage when there is white phlegm and stuffiness with congestion.

Nat. Mur. Running, watery colds with chilliness and a general feeling of discomfort. Loss of taste and smell and dryness of the skin.

Calc. Phos. A short course of this remedy is helpful in building up the general health after a cold.

COLIC

Colic is an attack of spasmodic pain in the abdomen attended usually by constipation. There are various forms of which simple colic is generally the result of the presence of undigested substances in the alimentary canal, which contracts spasmodically in an attempt to remove them. Pressure over the abdomen may bring relief and the anti-spasmodic tissue-salt, *Mag. Phos.*, should be taken with a sip of hot water every ten minutes. Infants are subject to attacks of colic, especially when fed artificially, and in such cases a modification of diet may be necessary.

Mag. Phos. With a sip of hot water every ten minutes until the pain is relieved.

Nat. Sulph. A useful supplementary remedy.

CONSTIPATION

Though persons in health generally have one daily movement of the bowels, some may have two regular motions, while in others a motion once in two days is quite normal. When the bowel is evacuated too seldom or incompletely, the motions become dry and hard with difficulty of evacuation. There may be a mechanical obstruction of the bowel, but generally the condition is due to one or more of the following factors—too little "roughage" in the diet, insufficient exercise, failure to inculcate the habit of regularity, dryness of the bowel, lack of tone of the colon muscle and of the villi of the intestines, etc. The diet should include plenty of fresh fruit, green vegetables, green salads (uncooked), whole-meal bread and dried fruits such as prunes, which have an indigestible residue and thus provide bulk to exercise the intestinal muscles. Drink plenty of fluids, including raw fruit juices which provide water in its purest form. The regular use of aperients tends to aggravate the trouble.

BIOCHEMIC TREATMENT

Kali Mur. When accompanied by indigestion with white-coated tongue and after excess of rich foods.

Nat. Mur. For constipation arising from dryness of the bowel.

Calc. Fluor. For relaxed condition of the bowel.

Nat. Phos. For chronic constipation in alternation with *Nat. Sulph.*

CONVALESCENCE

After an acute illness, and before complete health and strength are regained, the body requires a period of comparative rest in order to recuperate. Some acute ailments are attended by greater risks of a relapse during convalescence, and this applies particularly to those affecting respiration. During the period of recovery strenuous activities should be avoided and exposure to cold, damp, long hours of standing, etc., reduced to a minimum.

BIOCHEMIC TREATMENT

Calc. Phos. The principal remedy to restore the quality of the blood, to aid assimilation and to tone up the system generally.

Ferr. Phos. In alternation with Calc. Phos. to oxygenise the blood.

COUGHS

Coughing is a symptom that occurs during the course of most diseases of the respiratory organs. It should not, however, be neglected as it may be the forerunner of more serious trouble, e.g. bronchitis, pneumonia, etc. Expectoration varies in character according to the site in which it is produced, and the disease with which it is associated. Its nature and consistency is a guide to the tissue-salt required. Medical advice should be sought in chronic cases.

BIOCHEMIC TREATMENT

Ferr. Phos. Hard, dry cough with soreness and feverishness.

Kali Mur. Cough with white, albuminous phlegm; white or grey-coated tongue. Children's cough.

Kali Sulph. Cough with yellow expectoration. Worse in a heated room or in the evening.

Mag. Phos. Painful, spasmodic cough with a tendency to persist.

Calc. Sulph. When the cough is loose and rattling with expectoration of thin, watery sputum. In alternation with *Ferr. Phos.*

Silica. When the cough is accompanied by thick, yellow-green, profuse expectoration.

Calc. Phos. Useful as an intercurrent remedy and during convalescence.

CRAMPS

Cramp is a painful spasmodic contraction of muscles, generally affecting the limbs but sometimes the internal organs. It belongs to a group of ailments known as local spasms and treatment with the anti-spasmodic tissue-salt *Mag. Phos.* is indicated. The cause of these painful spasms is to be found in the nervous system. Cramp frequently comes on at night and during the attack the muscular fibres can be felt gathered up into a distinct knot. The attack is usually of short duration; massage and stretching the limb, by pressing against a firm object, are helpful remedial measures.

BIOCHEMIC TREATMENT

Mag. Phos. The principal remedy in cramps, spasms, neuralgias, twitchings, paroxysms, etc. Brings quicker relief when taken with a little hot water.

Calc. Phos. In alternation with *Mag. Phos.* Sensation as if parts were asleep, and with feeling of numbness and cold.

Silica. A useful alternative remedy if *Mag. Phos.* does not relieve.

CROUP

Croup is a term used to denote various ailments characterised by swelling which partially blocks the entrance to the larynx. It occurs in children and is accompanied by wheezing inspiration. It can sometimes be serious on account of the risk of a complete blockage and suffocation. The attack usually comes on suddenly at night, following a chill. The breathing is hoarse and croaking and there may be a struggling for breath. A child subject to attacks of croup should be especially guarded against cold and damp until the tendency is outgrown. Medical advice should be sought.

BIOCHEMIC TREATMENT

Ferr. Phos. Every fifteen minutes from the onset of the attack, in alternation with *Kali Mur.*

Kali Mur. The principal remedy for the exudation and swelling.

Mag. Phos. Spasmodic closure of the windpipe.

Calc. Phos. Useful if other remedies fail to bring relief.

DIARRHŒA

Diarrhœa is a symptom of many diseases and is one of the body's methods of ridding itself of unwanted substances. The checking of diarrhœa prematurely may therefore hamper instead of help the healing process.

Diarrhœa in infants may be due to gastro-enteritis, a serious condition calling for prompt medical attention and careful nursing. A predisposing cause is artificial feeding; environment and seasonal factors should be taken into account.

Among the many causes of diarrhœa in adults are indigestible substances and other errors in the diet, a catarrhal condition of the alimentary tract, emotional upsets, stomach chills, the eating of unripe fruits, food poisoning, unwashed greenstuffs, unaccustomed foods and alcoholic drinks, particularly those indulged in during holidays overseas. Rest and warmth will give nature a chance to deal with the trouble in her own way. A simple diet and the use of the appropriate tissue-salts will speed recovery.

BIOCHEMICAL TREATMENT

Ferr. Phos. Helpful in most cases of diarrhœa, more particularly where children are the subjects.

Nat. Phos. Diarrhœa with sour-smelling, greenish stools. Diarrhœa of teething children, often associated with worms. Yellow or creamy coating at the back part of the tongue. Summer diarrhœa with gastric weakness.

Calc. Phos. Diarrhœa resulting from digestive disturbances. *Calc. Phos., Ferr. Phos.* and *Nat. Phos.* cover most forms or diarrhœa of infants resulting from faulty assimilation of food.

Kali Mur. Diarrhœa caused by rich or fatty foods, pastries, etc. Pale stools and a white-coated tongue are indications for this tissue-salt.

Kali Phos. Diarrhœa caused by fright and other emotional upsets. Should always be considered for nervous subjects.

Nat. Sulph. Bilious diarrhœa. Dark-coloured stools. Worse in cold, damp weather and in humid conditions.

Nat. Mur. Diarrhœa alternating with constipation. Watery stools causing soreness and smarting.

Mag. Phos. As a supplementary remedy when the diarrhœa is accompanied by cramping pains and flatulence.

Calc. Sulph. Frequent, gushing stools giving the patient no rest. Give also *Ferr. Phos.* and *Calc. Phos.*

EARACHE

Earache may be due to many causes such as catarrh, boils and as a complication during fevers. In serious cases suppuration may occur and medical advice should always be sought if the symptoms persist. Wax in the ear is a common cause of deafness and it may be removed by syringing.

BIOCHEMIC TREATMENT

Ferr. Phos.   Inflammatory earache with burning, throbbing pain, after exposure to cold or wet.

Kali Mur.   Earache with swelling of the eustachian tubes. Catarrhal inflammation of the middle ear.

Nat. Mur.   Roaring in the ears, dullness of hearing with watery symptoms.

Kali Phos.   Dullness of hearing with noises in the head and accompanying nervous symptoms. Humming in the ears, especially in old people.

Calc. Sulph.   Discharges from the ear, sometimes mixed with blood.

Calc. Phos.   Discharge from the ears especially in children; the bones around the ear ache and there may be some swelling.

Mag. Phos.   Dullness of hearing from disorders of the auditory nerve fibres. Nervous earache.

Kali Sulph.   Earache with yellow discharge. Catarrh of the ear. An excellent remedy for the distressing "earache" of children.

EYES

Middle age is the time when eyesight tends to show signs of deterioration, especially for close work. When difficulty begins to be experienced in reading small print, that is the time to seek expert advice and a visit to the oculist is a wise precaution. Headaches can sometimes be traced to eye strain.

BIOCHEMIC TREATMENT

Ferr. Phos.   Inflammation without secretion of pus.

41

Burning with sensation as if grains of sand were under the eyelids. Conjunctivitis.

Kali Mur. Inflammation of the eye with whitish discharge.

Nat. Mur. Neuralgic pains with flow of tears.

Nat. Phos. Discharge of golden-creamy matter. Eyelids glued together in the morning. Acid symptoms.

Silica. Stye on the eyelid. Alternate with *Ferr. Phos.* for the inflammation.

FEVER

Fever is a condition characterised by an increase in body temperature. It is one of the most common symptoms of disease but should be regarded as secondary to the disordered state with which it is associated. The temperature of the human body in health ranges between 98·4 degrees and 99·5 degrees Fahrenheit. There are daily variations, the lowest being between the hours of midnight and 6 a.m. and the highest during the evening. The body normally maintains an even temperature by controlling the balance between heat gained and lost. Heat is gained by oxidation of the tissues (*Ferr. Phos.*) which takes place during the process of nutrition. Cooling occurs principally through the lungs and skin. In the feverish state balance is no longer maintained, more heat being lost than gained, the loss to the system being chiefly in the form of nitrogen compounds. A fever is usually preceded by chilliness and there may be headache or a feeling of tiredness in the limbs. There follows a "hot" stage, the skin feels dry, there is an increase in the rate of the pulse, excessive thirst, and little desire for food. Then follows some form of discharge, usually a profuse perspiration, after which the fever declines. In some diseases the fever may be continued or remittent. Rest in bed and warmth will assist the course of the fever and the aim should be to help the process of elimination, never to suppress it.

BIOCHEMIC TREATMENT

Ferr. Phos. The principal remedy for high temperatures, quickened pulse and feverishness.

Kali Sulph. In alternation with *Ferr. Phos.*, to control the temperature and to promote perspiration.

Kali Phos. Nervous fevers, high temperature, quick and irregular pulse with general nervous excitement.

Kali Mur. Catarrhal fevers, great chilliness, with white-coated tongue and constipation.

Nat. Mur. Hay fever with watery discharge; dryness of the bowel or other symptoms pointing to a disturbance in the moisture regulating processes.

FIBROSITIS

Fibrositis is the popular term applied to muscular rheumatism. Many different names have been given to the various manifestations of this condition—bursitis, myalgia, neuritis, lumbago, etc. The onset may be acute or gradual. It can affect a whole area, such as back and shoulders, or be localised in one place. Exposure to damp and cold is a predisposing factor. Fibrositis comes under the general group of ailments arising as a result of self-poisoning. Diet is accordingly important, and a course of treatment with the tissue-salts associated with the cleansing processes (*Nat. Sulph.* and *Kali Mur.*) should prove helpful.

BIOCHEMIC TREATMENT

Ferr. Phos. The principal remedy for acute, inflammatory pains brought on suddenly by chills, exposure, unaccustomed exercise, strains, etc.

Nat. Phos. In alternation with the principal remedy in cases of acidity.

43

Kali Mur. A useful supplementary remedy in acute cases.

Kali Sulph. When the pains are fleeting or shifting.

Nat. Sulph. To help in eliminating poison-charged fluids.

GASTRIC DISTURBANCES

Gastric disturbances include a wide range of ailments characterised by painful or uncomfortable symptoms associated with the function of digestion. The symptoms are numerous and may arise from simple causes such as too hurried meals and insufficient mastication of the food.

Diet is obviously important and the digestive organs should be given as much rest as possible by adopting regular habits and by the selection of wholesome, easily digestible foods. Acid dyspepsia, catarrh of the stomach and nervous dyspepsia are some of the conditions that come under this general heading. Medical advice should be sought in chronic cases.

BIOCHEMIC TREATMENT

Ferr. Phos. Gastritis with pain, swelling, and tenderness of the stomach. Dyspepsia with hot, flushed face. Vomiting of undigested food, the tongue being clean.

Kali Mur. Gastric derangements when the tongue has a white or greyish-white coating. Indigestion or nausea after taking fatty or rich foods. In alternation with *Ferr. Phos.* in the treatment of gastritis.

Nat. Phos. Gastric derangements with symptoms of acidity. Sour risings, creamy, golden-yellow coating on the back part of the tongue. Heartburn after eating. Fretful, irritable disposition.

Calc. Phos. A useful remedy in the treatment of gastric and digestive disturbances when taken in alternation

44

with other indicated remedies. It aids the digestive processes and improves assimilation.

Kali Phos. Nervous indigestion with "gone" sensation in the stomach. Hungry feeling after taking food. Stomach-ache after fright or from excitement.

Kali Sulph. Gastric catarrh with slimy golden-yellow coating of the tongue. Colicky pains in the stomach with feeling of pressure or fullness.

Mag. Phos. Spasmodic pains and cramp of the stomach, relieved by hot drinks. Flatulence with much belching of gas.

Nat. Sulph. Gastric disturbances with bilious symptoms, bitter taste in the mouth, vomiting of bitter fluids. The tongue is sometimes coated a greenish-brown or greenish-grey colour.

HÆMORRHAGE

Hæmorrhage is an unusual flow of blood from any part of the body, caused by injuries, weakness of the blood vessels, etc. It may be external or internal; from an artery, when the blood is bright red and comes in spurts, corresponding to the heart beats, or from a vein, when it is of a darker colour and wells up into the wound. Hæmorrhage calls for immediate medical attention and until help arrives the usual first-aid steps should be taken to prevent loss of blood. The patient should be kept warm and not disturbed more than necessary. Slight hæmorrhages, such as occur as a result of piles or prolonged and excessive menstruation, may lead to an anæmic condition of the blood (*Calc. Phos.*). A little powdered *Ferr. Phos.* applied direct to the injured parts will help to stop the bleeding and aid the wound to heal cleanly.

BIOCHEMIC TREATMENT

Ferr. Phos. Bleeding from wounds; bleeding into the tisues as in bruises, etc., if possible, apply powdered *Ferr. Phos.* locally.

HÆMORRHOIDS

Hæmorrhoids (piles) consist of a varicosed condition of the veins at the lower end of the bowel. Piles may be internal or external, or both. Internal piles extend about one inch up the bowel. Sedentary habits are a predisposing factor and constipation with straining at stool is not an unusual accompaniment of this painful and embarrassing condition. Piles sometimes occur during pregnancy and they can also be a symptom of other diseases. External piles need not necessarily cause any pain or trouble. Sometimes they may become inflamed and give off a thin discharge. Internal piles may give no sign of their presence except for occasional bleeding. Too great a loss of blood will cause anæmia. When internal piles are large they may protrude and become inflamed and painful. In general piles are more of an inconvenience than a serious condition. Care should be taken to avoid constipation and it is better to achieve this by regulating the diet than by recourse to purgatives, which in the long run aggravate the trouble. The parts should be carefully washed after evacuation and any protrusions gently replaced. Piles tend to be one of the ailments of middle-age. Medical advice should be sought.

BIOCHEMIC TREATMENT

Ferr. Phos. For the inflammation and bleeding.

Calc. Fluor. Internal and blind piles, sometimes accompanied by pain in the back. Tones up the relaxed condition of the veins and muscular fibres.

Calc. Phos. Intercurrently with *Calc. Fluor.* in anæmic people.

See also Combination G.

HAIR, FALLING OUT

Every day, in the healthy scalp, a certain number of hairs reach the end of their existence, and are combed out, being replaced in time by new hairs. Each follicle produces many hairs in the course of a lifetime, but sometimes the hairs become gradually finer and the hair producing quality of the follicle finally fails. This may be due to an eczematous condition of the scalp, or dandruff, and to a certain extent it may be hereditary. Some diseases may cause partial loss of hair and prolonged anxiety or nervous shock are known to be causative factors. A little castor oil massaged into the scalp with the finger tips is helpful.

BIOCHEMIC TREATMENT
Kali Sulph. Falling-out of hair, bald spots. Much scaling of the scalp, moist and sticky. Dandruff.
Silica. Impoverished condition of the hair, lack lustre. A valuable hair conditioner.
Nat. Mur. A useful supplementary remedy in cases of dandruff and falling-out of hair.

HAY FEVER

Hay fever generally occurs during the summer months in people who are susceptible to pollen, or some irritating substance. This leads to the production of too much histamine which is responsible for the symptoms. It begins with an itching of the eyes and nose, followed by symptoms of a cold. There may also be headache, severe sneezing, and a watery discharge from eyes and nose. Precautions should be taken to avoid the exciting cause, for example, meadows during the summer season in the case of a person affected by pollen from hay fields. *Nat. Mur.* is the tissue-salt usually indicated.

BIOCHEMIC TREATMENT
Nat. Mur. For hay fever after exposure to sun; watery symptoms with sensation of itching and tingling in the nose.

47

Mag. Phos. To prevent a threatened attack from maturing, or to relax the spasms.

Silica. Itching and tingling of the nose with violent sneezing.

Kali Phos. For the depression and to aid breathing.

Ferr. Phos. For the congestion, inflammation and headache.

HEADACHE

Headache is one of the most commonplace of human ailments. It is symptomatic of many diseased conditions, some slight and others more serious, and it can arise from a variety of causes. The brain itself is insensitive to pain. It is probable that the pain of headache is due to dilatation of the arteries or to pressure of some sort. The dilatation may be a response to nervous stimuli, or the pain may arise from the neck muscles, or scalp, or from other nearby organs. One of the most frequent causes of headache is anxiety and living at too high a pressure; overworked professional people and business executives are likely subjects.

Defective eyesight is another common cause. The pain occurs in the region of the brow and tends to come on in the evening, particularly if a lot of reading or close work has been done during the day. Sinus infection is a less common cause and in such cases there is usually a history of colds in the head. Teeth can also be responsible for headaches. Indigestion accompanied by nausea may bring on the type of headache known as migraine. Constipation is a common cause, especially in children. Headaches are also a characteristic feature of fevers.

These are some of the factors responsible for headaches and in treating the trouble it is necessary to search out and remove the cause. Although the pain-killing drugs have their place, their continued use to suppress symptoms is not advisable and can result in harmful side-effects, e.g. stomach hæmorrhage. When headaches occur regularly the proper course is to consult a doctor and to put him in possession of as much information as possible in order to help him to ascertain the cause. Applications of eau-de-cologne to the forehead are sometimes helpful and the tissue-salts are useful as first-aids, e.g. *Ferr. Phos.* for the congested, throbbing headaches associated with the dilatation of the arteries.

BIOCHEMIC TREATMENT

Ferr. Phos. Inflammatory headaches from cold, sun-heat, with throbbing in the temples or over the eyes. Congestive headache at the menstrual period. Soreness of the scalp. Pain worse from motion or noise.

Kali Phos. Nervous headaches in alternation with *Mag. Phos.*

Kali Mur. Headache with white-coated tongue, disordered stomach and sluggish liver.

Kali Sulph. Headache which is worse in a heated room and in the evening.

Nat. Mur. Dull, heavy headache, with drowsiness, sleep not refreshing.

Nat. Sulph. Sick headache with biliousness; nausea and bitter taste in the mouth. Worse in damp, warm weather. Migraine.

Nat. Phos. Headache on the crown of the head on awakening. Sick headache with acid symptoms, especially after taking wine or milk.

Calc. Phos. Headache with vertigo and in anæmic persons.

Calc. Sulph. Headache with vertigo and nausea. Pain around the whole head.

HICCUPS

Hiccups is a spasmodic contraction of the muscles of the diaphragm closing the throat, resulting in a sudden

49

shutting off of breath. It is brought about by an irritation of the nerves which serve the diaphragm, usually following some digestive upset. Overeating, or too rapid eating, sometimes brings on an attack, especially in children.

BIOCHEMIC TREATMENT

Mag. Phos. The principal remedy. It should be given with a little hot water.

Nat. Mur. For hiccups after hasty eating. In alternation with *Mag. Phos.* if the spasms persist.

HIVES (Nettlerash)

Hives is another name for nettlerash. It is a skin eruption somewhat resembling the effect produced by the sting of nettles. There is considerable itching and irritation which may extend over large areas of the body. It is usually brought on by digestive disturbances or after eating certain kinds of foods, such as shellfish. It is one of those ailments where the sufferer is allergic to the exciting cause and that should be sought out and avoided. In some cases it is accompanied by feverishness and the eruptions can cause temporary swelling and disfigurement and there may be other toxic symptoms. The anti-toxic tissue-salts, *Kali Sulph.* and *Nat. Mur.* are helpful in treating this trouble.

BIOCHEMIC TREATMENT

Ferr. Phos. When there is feverishness.

Kali Sulph. When the skin is dry and tending to scale.

Nat. Mur. Eruptions with clear, watery contents, nettlerash after becoming overheated.

Nat. Phos. Soreness of the skin, with symptoms of acidity. Creamy exudations.

INDIGESTION (See Gastric disturbances)

INFLUENZA

Influenza is one of the infectious, febrile diseases principally involving the respiratory organs. It occurs usually in epidemics during the winter months. The Italians at one time ascribed it to the influence of the stars, hence the name "influenza". It is now known to be due to a virus, of which there are three types—A, B and C. Type "A" is the most prevalent in this country. The onset of influenza is sudden with a feeling of chilliness, headache and aching of the limbs, followed by sore throat and other symptoms. Old people, particularly, are susceptible to complications, e.g. pneumonia, and the chief mortality of influenza is due to such complications. A relapse can occur if the person tries to return to normal duties prematurely so extra care should be taken during convalescence.

BIOCHEMIC TREATMENT

Nat. Sulph. The principal remedy. It should be taken in alternation with *Ferr. Phos.* throughout the feverish stage.

Ferr. Phos. For the inflammation and feverishness.

Kali Sulph. To promote perspiration and to control temperature.

Kali Mur. For the catarrhal symptoms in alternation with one or more of the other indicated remedies.

Calc. Phos. During convalescence.

LUMBAGO

Lumbago is a form of rheumatism affecting the muscles of the lower part of the back. It is sometimes brought on by exposure to cold and damp or unaccustomed exercise, such as bending and lifting. There is an inflammatory condition of the muscular tissues (*Ferr. Phos.*) with congestion and pressure upon the nerve endings. Lumbago usually comes on suddenly, like a

51

stab in the back, and it may be difficult or impossible to move on account of the pain. The attacks are usually of short duration. Treatment is on similar lines to that recommended for rheumatism. The local application of heat and gentle massage, if this can be borne, should bring a measure of relief.

BIOCHEMIC TREATMENT

Ferr. Phos. In the early stages for the inflammation and pain.

Nat. Phos. To counteract acid conditions.

Nat. Sulph. In alternation with *Nat. Phos.* in acid subjects.

Calc. Fluor. Lumbago following a strain, in alternation with *Ferr. Phos.*

Calc. Phos. Severe pains on bending, unable to straighten. Rheumatic pains in the joints with cold or numb feeling. *Calc. Phos.* symptoms are usually worse from cold and change of weather.

MALARIA

Malaria is a disease caused by the presence of parasites in the blood, following a bite from a certain species of mosquito. It has been known from earliest times and its connection with damp, swampy surroundings has long been recognised. Malaria is chiefly confined to tropical climates, but is not unknown in temperate regions. The parasite is carried by the Anopheles mosquito which breeds in the surface water of pools and in areas of rank vegetation.

The acute malarial attack has three stages, chill, fever and sweat, the paroxysms recurring at twenty-four, forty-eight and seventy-two hours, according to the type of malaria. There is also a form known as remittent fever in which the temperature does not become normal for from one to two weeks, with much muscular soreness and aching. Malaria is prevalent where humidity causes an excess of water in the blood, excluding the proper amount of oxygen from the system. The paroxysms disappear with the elimination of the excess water, coming on again as more water is absorbed from the moisture-laden atmosphere. The chances of an attack of malaria can be lessened by making sure that the system is plentifully supplied with the water-eliminating tissue-salt, *Nat. Sulph.*

BIOCHEMIC TREATMENT

Ferr. Phos. An important remedy for the fever, especially when there is vomiting of undigested food.

Nat. Sulph. The principal remedy in all cases; bilious stools, dirty grey-green coating of the tongue, bitter vomiting.

Calc. Phos. For the anæmic condition following an attack.

MEASLES

Measles is an acute and highly infectious disease occurring mostly in children. The disease known as German measles is a milder form but special care should be taken by women during pregnancy to avoid contact with infected persons. Epidemics usually occur during the winter months and it is believed that the infecting agent is a filterable virus. Measles usually begins like an ordinary cold, with cough, watering at the eyes and nose and a high temperature. In four or five days eruptions appear, small red spots, beginning on the face and neck, gradually running together and extending down the body. The room should be kept darkened if the eyes are affected and the child given a jig-saw puzzle or game in prefence to reading. About the ninth day the eruption disappears, with a bran-like shedding of the outer skin. Measles is contagious and it is possible for infection to be carried from one place to another by clothing or other articles or by an unaffected third person. There is always a risk of complications such as bronchitis or pneumonia and special care should be taken during convalescence. Medical advice should be sought.

BIOCHEMIC TREATMENT

Ferr. Phos. The first remedy for the fever and inflammation.

Kali Mur. The second remedy for the cough, swelling and coated tongue.

Kali Sulph. When the rash appears and for the peeling of the skin.

Calc. Phos. When the skin symptoms have cleared up and during convalescence.

MENSTRUATION

Menstruation is a normal function beginning at the age of puberty and ending at the change of life. It consists of a periodic discharge of blood and mucus from the uterus. Menstruation occurs normally about every twenty-eight days, varying slightly at longer or shorter intervals in different women. The duration of the flow and the amount of blood lost also varies considerably but each woman finally establishes a standard which is normal to herself.

The regularity of menstruation depends upon the state of health and is easily upset by any changes in occupation, climate or surroundings. Any factor affecting the mind may bring about a temporary disturbance of menstruation.

Hygiene during the menstrual period should be the same as at any other time except that mental and physical exertions should be somewhat lessened. Moderate exercise, fresh air, nourishing food and plenty of sleep are needed.

Dysmenorrhœa is the name applied to painful symptoms which may be present during the menstrual period. It arises from various causes, congestion and inflammation of the pelvic organs, neuralgia from irritability of the nervous system, chills, anæmia, etc. Medical advice should be sought.

BIOCHEMIC TREATMENT

Ferr. Phos. Painful menstruation, bright red flow, flushed face, quickening pulse. Congestion of the pelvic organs with too profuse loss of blood.

Kali Mur. Menses too early and too frequent, or too late or checked from taking cold. Black, or dark clotted blood.

Mag. Phos. For spasmodic pains, cramp, labour-like, bearing-down pains, menstrual colic.

Kali Phos. Menstrual colic in nervous, sensitive women, Dark red flow.

Calc. Phos. Intercurrently in anæmic women and girls at puberty. Scanty flow.

Nat. Phos. When there are acid symptoms and for local irritation.

Nat. Mur. For depression of spirits and lassitude. Thin watery discharge. Headache in the mornings.

MUMPS

Mumps is an infectious disease characterised by an inflammatory swelling of one or other of the salivary glands, usually the parotid, and frequently occurring as an epidemic. It mostly affects young persons but can occur at any age. It is highly infectious for two or three days before the swellings appear. There is an incubation period of two to three weeks, after infection, before the glands begin to swell. The first signs are feverishness, sore throat, and high temperature. The swelling goes

55

down within about a week. The person should be kept in isolation for about fourteen days from the onset of the disease or seven days after the subsidence of all swelling. In some cases there are swellings in other glands of the body and care should be taken to guard against complications. Medical advice should be sought.

BIOCHEMIC TREATMENT
Ferr. Phos. During the first stage whilst the fever lasts, in alternation with *Kali Mur.*
Kali Mur. The principal remedy for the glandular swelling and pain on swallowing.
Calc. Phos. During convalescence to restore strength and vitality.

NERVOUS DEBILITY

This is a state where nerve force is being used up more quickly than it can be generated, and when we recognise this simple fact we realise that to stimulate the nerves with so-called "tonics" is one way of aggravating the trouble. The nutritional aspect of this condition must be borne in mind and those tissue-salts which play a part in elaborating the nerve fluids should be taken steadily over a reasonable period. Grief, worry, undue mental exertion, irregular habits, emotional extravagance; these are the parasitical influences which rob the nerves of their vitality. Treatment should aim to increase the supply of nervous energy and to bring the nerves back to a state of normal tranquillity. When this has been achieved a remarkable change for the better will be experienced.

BIOCHEMIC TREATMENT
Kali Phos. This is the principal nerve tissue-salt; the nerve nutrient and vitaliser which should be given in ailments of a nervous character. It is the remedy when the nerves are said to be "on edge".

Mag. Phos. This tissue-salt is another valuable nerve nutrient. It acts well with *Kali Phos.* but has more to do with the motor nerves. *Mag. Phos.* is indicated for nerve pains, cramps and nervous twitchings. It helps to steady the nerves.

Calc. Phos. This tissue-salt is needed to raise the general nutritional tone and to improve the quality of the blood. It promotes the assimilation of vital nutrients and so contributes to the supply of nervous energy. Other tissue-salts may be needed to deal with individual symptoms but the above are the most frequently needed remedies for ailments of a truly nervous character.

NEURALGIA

Neuralgia is a nerve pain, following exposure to cold, injury, fever, pressure, irritation, etc. These pains are most common in the nerves of the face and head, the symptoms consisting of shooting and intense pains along the course of the nerves. Attacks of neuralgia are apt to occur when the general health is in a low state and when this is suspected a course of *Calc. Phos.* will prove helpful.

BIOCHEMIC TREATMENT

Mag. Phos. The principal remedy for neuralgic pains.

Kali Phos. Neuralgic pains in nervous persons. For the depression, sleeplessness and irritability.

Ferr. Phos. Acute neuralgic pains due to inflammatory conditions, caused by chills, fevers, etc. Alternate with *Mag. Phos.*

Calc. Phos. To improve the general state of health.

NEURITIS

Neuritis is an inflammatory condition of a nerve or nerves which may be localised in one part of the body

57

or may be general. The fibrous sheath of the nerve may be irritated by inflammation (*Ferr. Phos.*) cold, pressure, or some other cause. The symptoms vary according to the location of the trouble and certain forms of neuritis are associated with a vitamin deficiency.

<small>BIOCHEMIC TREATMENT</small>

Ferr. Phos. For the inflammatory condition.

Kali Phos. Intercurrently to improve the nutritional tone of the nerves.

Mag. Phos. For the spasmodic pains.

Silica. A useful supplementary remedy for the nerve sheath tissues.

PAIN

Pain is a timely indication that the human machine is not running as smoothly as it should; there is friction somewhere. It has been well said that "*pain is a prayer of a nerve for relief.*" Give prompt heed to these pains, note their nature and location and select the tissue-salt most closely corresponding, e.g. *Ferr. Phos.* throbbing pains, *Mag. Phos.* cramping pains, etc.

<small>BIOCHEMIC TREATMENT</small>

Ferr. Phos. Throbbing pains with heat, inflammation and congestion, strained muscles and tendons, sprains. Pains resulting from cuts and wounds (a little powdered *Ferr. Phos.* should be applied locally). *Ferr. Phos.* and *Mag. Phos.* are probably the two most frequently needed remedies in the treatment of pain.

Mag. Phos. Spasmodic, cramping pains, with acute stabbing, boring sensation, neuralgia, sciatica, menstrual pains, etc.

Kali Phos. Itching of the skin with nervous irritation or crawling sensation, chilblains which itch and tingle.

Calc. Fluor. Aches and pains of the legs with feeling

of heaviness due to bad circulation. Pains in the lower part of the back with dragging sensation.

Calc. Phos. Deep-seated pains in the bones and joints, severe at night, with sensation of numbness or the trickling of cold water. The pains are alleviated by moving the limbs.

Kali Mur. Pains accompanied by soft swellings, face-ache with swelling of the cheeks or gums, tonsillitis, etc. Pains accompanied by white, fibrinous discharges. Gastric pains after eating fatty or rich foods. This tissue-salt is the first-aid for burns.

Kali Sulph. Stomach pains with sensation of pressure and fullness at the pit of the stomach. Pains which are worse in the evening, or in a heated, stuffy atmosphere. Shifting twinges of pain. Alternate with *Ferr. Phos.* in the treatment of inflammatory pains to promote perspiration when the skin is hot, dry and harsh.

Nat. Mur. Pains accompanied by an increase of watery secretions such as tears, nasal discharges and urine. Pulsations felt all over the body. Pains in the back relieved by lying on something hard. Painful blisters and blebs on the skin with watery contents. Itching of nettle-rash. The pains sometimes make their appearance at regular intervals.

Nat. Phos. Digestive pains when associated with acidity, heartburn, sour acid risings, etc. Itching of the nose and anus may also be symptoms of an over-acid state of the blood.

Nat. Sulph. Pains associated with liver disturbances, notably biliousness, sick headache, nausea, etc.

Silica. Pains due to the formation of boils, gumboils, abscesses, etc. Suppurating pains and festering conditions call for the deep-acting, eliminating function of this tissue-salt. Pains arising from neglected injuries with suppuration.

PILES (See Hæmorrhoids)

RHEUMATISM

Rheumatism is not a single ailment; the term embraces a whole group of disabilities arising from many and varied causes, among which is auto-intoxication or self-poisoning.

At the onset there is a measure of congestion and inflammation and if this can be broken up promptly a serious attack may be averted. For this purpose the tissue-salt *Ferr. Phos.* is needed. *Ferr. Phos.* is the oxygen-carrier of the blood, its action enables the tissues to "breathe" and so to burn up their waste products. The local congestion is thus relieved and the inflammation subsides. *Ferr. Phos.* is useful as a first-aid for acute attacks of rheumatism.

Another associated symptom of rheumatism is acidity. Faulty elimination allows the accumulation in the blood of acid-waste products which have a bad effect on the general circulation. The acid-neutralising tissue-salt, *Nat. Phos.*, breaks up these harmful acids and so helps to bring about their elimination.

In rheumatic ailments, all the eliminative organs are involved. Kidneys, liver, bowels, lungs—even the skin. There is some hold-up in the body's waste transport system, the organs concerned are lacking in tone and efficiency. The great vitaliser of this transport system is *Nat. Sulph.* This tissue-salt brings about the removal of the poison charged fluids which are the normal result of the chemical exchanges constantly taking place in the living tissue-cells. If conditions arise which allow these waste matters to accumulate in the blood and tissues, then auto-intoxication (self-poisoning) results. The importance of *Nat. Sulph.* to rheumatic subjects thus becomes self-evident.

Other tissue-salts may sometimes be required owing to complicating conditions, but the three tissue-salts mentioned are the ones most frequently needed.

Ferr. Phos. For the pain, inflammation and congestion.

Nat. Phos. To neutralise the acid-forming tendency.

Nat. Sulph. To aid in removing toxic-charged fluids from the system.

Silica. To break up accumulation of urates lodging around the joints and muscles.

SCIATICA

Sciatica is inflammation of the great sciatic nerve which runs down the back of the thigh. It can follow as a result of exposure to cold and damp, causing irritation of the nerve. There may be a rheumatic tendency. Pressure from other causes may also be responsible for sciatic pain. A slipped vertebral disc is not an uncommon cause of pain in the sciatic nerve. The trouble may first be felt a little behind the hip joint, and then extending downwards even as far as the foot. Hip, knee and ankle joints are particularly tender areas. In severe cases movement of the limb is very painful and rest in bed becomes necessary. For the pain and inflammation the chief remedy is *Ferr. Phos.* taken every half-hour during an acute attack.

BIOCHEMIC TREATMENT

Ferr. Phos. For the general pain and inflammation.

Mag. Phos. When the pain is spasmodic.

Kali Phos. In alternation with *Mag. Phos.* when there is nervous exhaustion with great restlessness.

Nat. Sulph. Pain when getting up from sitting or turning in bed, no relief in any position.

SINUS

Sinus is a cavity in bone or tissue. The air sinuses of the frontal bones communicate with the interior of the nose. Inflammation or infection may spread into the sinus cavities and may be difficult to disperse on account of the restricted drainage apertures. Suppuration into the nasal sinuses may be associated with an abscess of the upper teeth the roots of which project into the cavity. Nasal catarrh may also spread infection into the sinus.

BIOCHEMIC TREATMENT

Ferr. Phos. For the inflammation (sinusitis) in alternation with the remedy indicated by the nature of the discharge.

Calc. Phos. Albuminous discharge, like the white of raw egg.

Nat. Mur. Clear, watery discharge causing soreness. Salty taste.

Kali Mur. White, fibrinous discharge.

Kali Sulph. Sticky, yellow or greenish discharge.

Calc. Fluor. Yellow, lumpy discharge affecting the bones. Alternate with *Silica.*

Silica. Purulent discharge, alternate with *Calc. Sulph.* Where the bone is affected alternate with *Calc. Fluor.*

SKIN AILMENTS

Skin ailments are a large and important group which not only influence the general health, but may reveal constitutional weaknesses both inherited and acquired. There are several broad classifications, those affecting the sebaceous and sweat glands, inflammatory diseases, nervous disorders, and those due to parasitic infections. Medical advice should always be sought in the first instance to determine the nature and cause of the trouble.

It should be borne in mind that the skin is an important organ of elimination and that most forms of discharge are indications that the system is throwing-off unwanted organic materials which for some reason have become non-functional.

Seborrhœa is a term applied to an accumulation of sebaceous secretions forming scales mostly on the scalp and which may interfere with the growth of the hair. Acne, is a somewhat similar form of eruption occurring more on the face and upper part of the body, and especially during adolescence. Careful cleansing of the affected parts is necessary. A wart (verruca) is an excrescence from the surface of the skin, which sometimes disappears spontaneously or may have to be excised. Inflammatory affections are symptomatic of many fevers and may take various forms. Nettlerash is a diffuse redness of the skin accompanied by weals similar to the effect of stinging nettles.

Catarrhal conditions of the skin are a large and important group, the most common being eczema, which may be wet or dry. Shingles is a painful eruption which may attack any part of the body but always along the distribution of a nerve. Outward applications of powdered *Ferr. Phos.* help to relieve the pain. Shingles is more frequently met with in elderly people and it can be a very debilitating ailment.

As a general rule, the symptomatic picture will indicate the tissue-salt required—inflammation (*Ferr. Phos.*), scaling (*Kali Sulph.*), whitish discharges (*Kali Mur.*), watery or dry states (*Nat. Mur.*), and so on. Outward applications are helpful. The diet should be wholesome and nourishing and should include plenty of fresh fruits and green vegetables. Avoid highly-seasoned dishes and over-refined foods such as white bread, pastries, sweets, white sugar, etc.

BIOCHEMIC TREATMENT

Ferr. Phos. In the first stage for the inflammation, heat, burning, pain, etc. Outward application in powdered form for the pain of shingles.

Kali Mur. Second stage of inflammatory conditions.

Eruptions are thick and white, and may be accompanied by white-coated tongue and light coloured stools. Warts, shingles, etc., in alternation with *Nat. Mur.*

Kali Sulph. Skin eruptions which are sticky, yellow and watery. Dry skin with suppression of eruptions; peeling of the skin with or without sticky secretions. Symptoms are aggravated in the evening and in hot, stuffy atmospheres. The principal remedy for relief from the effects of psoriasis.

Nat. Mur. Skin eruptions when the discharges are clear and watery. Excessive dryness of the skin. Blisters, nettlerash, bites and stings of insects (applied locally). Shingles in alternation with *Kali Mur.*

Kali Phos. An important constitutional remedy when the trouble is of nervous origin, e.g. shingles. Great irritation of the skin; offensive discharges.

Silica. Abscesses, boils, styes, etc. Thick, yellow discharge. Suppressed perspiration.

Calc. Phos. Pimples on the faces of young persons in alternation with *Calc. Sulph.* Pallid, anæmic appearance.

Calc. Fluor. Chaps and cracks of the skin; cracks in the palms of the hands.

Calc. Sulph. Pimples during adolescence, acne. A useful intercurrent remedy to cleanse the blood stream and hasten healing.

SLEEPLESSNESS

Sleeplessness may be due to various causes. People of a highly-strung disposition find it more difficult to relax—nervous tension, aggravated by worry, anxiety, overwork, etc. is one of the more common factors. Indigestion is responsible in many cases and the eating of a heavy meal and stimulants too soon before retiring should be avoided. The brain sometimes becomes over-active as a result of late night work so that it is diffi-cult to settle down to sleep. A certain amount of mental effort is called for to stem the kaleidoscope of thoughts that tend to crowd the mind. It is sometimes helpful to take a hot drink and read a chapter from a favourite book, thus breaking the recurring chain of thoughts. A dose of *Nat. Phos.* kept handy at the bedside will soothe minor digestive upsets; *Kali Phos.* is helpful as a con-stitutional remedy when the nervous system is run-down. A hot drink, taken slowly just before bedtime, is an effective nightcap, and adequate warmth, especially of the feet, and fresh air are aids to a sound night's sleep. Elderly people usually need less sleep and sleep-lessness at night can sometimes be made good by a nap after lunch. Medical advice should be sought when sleeplessness persists.

BIOCHEMIC TREATMENT
Nat. Phos. In alternation with *Nat. Sulph.* when due to digestive disturbances.
Kali Phos. Sleeplessness from nervous causes or due to over-excitement.

SORE THROAT

An inflammatory condition of the wall of the throat (the pharynx) is known as pharyngitis and inflammation of the mucous membrane of the larynx (the organ of the voice) is laryngitis. The term "throat" is popularly applied to the region about the front of the neck, but, correctly speaking, it means the irregular cavity into which the nose and mouth open above, and from which the larynx and gullet lead below, where the channel for the air and that for the food intersect.

Pharnygitis may be due to infection, digestive disturb-ances, too much smoking, etc. and in severe cases there may even be slight bleeding of the smaller blood vessels.

There is usually irritation, cough, and general discomfort. Laryngitis is brought on by similar causes including exposure to damp and draughts, too much talking, etc., and it is also a symptom of many infectious diseases. There may be swelling and difficulty in breathing. The heat, pain and dryness are a clear call for *Ferr. Phos.*, the oxygen-carrier, which should be given at frequent intervals during the acute stage until the inflammation subsides. A cold compress may help to relieve the pain and inflammation.

BIOCHEMIC TREATMENT

Ferr. Phos. For the inflammation and burning pain. Throat red and dry with hoarseness, loss of voice.

Kali Mur. In alternation with *Ferr. Phos.* when there is swelling of the glands or tonsils.

Calc. Sulph. If taken in the early stages can prevent development of a cold. In the later stages of tonsillitis when matter is discharged or abscess forms. Ulcerated sore throat.

Calc. Fluor. Relaxed throat with tickling in the larynx. In alternation with *Calc. Phos.* for enlargement of the tonsils.

Calc. Phos. Sore, aching throat with pain or swallowing. Chronic enlargement of tonsils.

SUNSTROKE

Sunstroke is caused by exposure to the sun or overheated air (heatstroke) especially in humid atmospheres. The body becomes overheated owing to a temporary disorganisation of the heat-regulating mechanism. There may be headache, lassitude, dizziness and temporary loss of consciousness. The blood pressure is low and there may be vomiting. The person should be placed in the shade, tight clothing loosened, but care taken to avoid chilling. A saline drink on recovery promotes sweating—a favourable reaction. The water-distribution in the tissue at the base of the brain, the actual cause of collapse, must be equalised and the tissue-salt for this purpose is *Nat. Mur.*

BIOCHEMIC TREATMENT

Nat. Mur. This is the chief remedy to regulate the distribution of moisture. It should be given at frequent intervals.

Ferr. Phos. In alternation with *Nat. Mur.* for the inflammatory symptoms and to help respiration.

SYNOVITIS

Synovitis is inflammation of the membrane lining a joint. There is usually an effusion of fluid with swelling and pain. It occurs in certain rheumatic diseases and also as a result of injuries and strains, particularly those arising from athletic activities. Bursitis is a somewhat similar condition.

BIOCHEMIC TREATMENT

Ferr. Phos. For the pain, stiffness and inflammation.

Nat. Sulph. To disperse the infiltration of fluid.

Silica. For chronic synovitis of the knee, with swelling and difficulty of movement.

Calc. Fluor. A useful remedy in long-standing cases that are slow to respond to treatment.

TEETH

Every part of the body is built up as a result of the chemical combination of inorganic tissue-salts with organic materials. The teeth, no less than bones, flesh and sinews, must be supplied with their requisite inorganic constituents if decay is to be prevented and a healthy state preserved.

One of the essential constituents for the formation of teeth is *Calc. Phos.* When this tissue-salt is deficient, dentition will be delayed or otherwise disturbed. Dentition begins normally at the fourth to sixth month in infancy, and the temporary set of milk teeth, as they are called, should be completed by the thirtieth month. The lower central teeth are the first to erupt followed by the upper central and filling in, in order, towards the back till the set of twenty is complete. During the sixth year the milk teeth begin to shed and the permanent teeth take their place. This set consists of thirty-two teeth and is not complete till the eighteenth to the twentieth year.

Infants frequently have constitutional disturbances during the period of dentition shown by increased irritability or more profound reactions such as diarrhœa, fever or even spasms.

The milk teeth should be cared for as carefully as the permanent ones, as the quality and position of the latter depend on the temporary teeth remaining sound and in place until pushed out by the eruption of the permanent set.

BIOCHEMIC TREATMENT

Calc. Phos. Teeth develop slowly and decay rapidly. This is the principal nutrition remedy for the teeth. Important for teething infants, children and expectant mothers.

Calc. Fluor. Enamel of the teeth rough and deficient causing rapid decay. Looseness of the teeth in their sockets. Delayed dentition, in alternation with *Calc. Phos.*

Ferr. Phos. Inflammatory toothache with soreness; bleeding after extractions, in alternation with *Kali Mur.* when there is swelling.

Kali Phos. Bleeding of the gums, severe pain in decayed or filled teeth.

Mag. Phos. Teeth very sensitive to touch or cold air; severe toothache with shooting pains.

Nat. Phos. Grinding of the teeth during sleep.

Silica. Gums painful on slight pressure, gumboils, abscess at the roots.

Calc. Sulph. Toothache, with inside of gums swollen and sore. Gums bleed on brushing teeth. In alternation with *Silica* for the treatment of gumboils and ulcerations.

TONSILLITIS

Tonsillitis is inflammation of the tonsils and may be either acute or chronic. In quinsy the condition is not entirely confined to the tonsils as there is involvement of the surrounding areas. Infection occurs mostly during the winter months. Predisposing factors are similar to those preceding the onset of seasonal coughs and colds. Tonsillitis can be infectious and in children it may be a forerunner of more serious trouble. The onset is sudden with pain in swallowing, chilliness and fever. The tonsils become enlarged and exude a whitish purulent substance (*Kali Mur.*) and the glands of the throat may become tender and swollen. Medical attention is needed as during the early stages the symptoms are similar to those of diphtheria. Rest in bed, with a light diet and plenty of fluids will help the body to overcome the infection. Eating is painful and difficult but with plenty of hot milk and fruit drinks there is no need to worry on that account and the stomach will be all the better for the respite from solid foods. There is always a risk of complications in this kind of infection (hæmolytic streptococcal) unless proper precautions are taken and this applies particularly with children. Keep them in bed for at least two days after the temperature has returned to normal. The tonsils play an important part in the protective mechanism of the body; they act in the manner of filters and any exudation shows that they are doing their job efficiently. Medical advice should be sought.

Ferr. Phos. The first remedy. Tonsils red and inflamed, painful on swallowing.

Kali Mur. The second remedy, as soon as there is any swelling. Tonsils spotted white or grey.

Silica. Periodical tonsillitis and when suppuration threatens.

Calc. Sulph. During the last stage, when purulent matter discharges.

Calc. Phos. Chronic swelling of the tonsils, causing pain on opening the mouth, with difficulty in swallowing.

Calc. Fluor. Large indurated tonsils, relaxed throat and elongation of the uvula.

URINARY SYMPTOMS

The urinary system comprises the kidneys, which filter the urine from the blood, two tubes called ureters, through which the urine flows from the kidneys to the bladder, whence it is voided through a further short tube called the urethra.

Urine consists chiefly of water in which are dissolved waste substances resulting from cellular activity. The amount of water lost daily by the body through perspiration varies with the season, as is also the case with the kidneys when more is passed in the winter than in the summer. Regarding the loss of water by the body, the glands producing perspiration and the kidneys are complementary to each other. Urine contains about 4 per cent solids—urea, common salt, phosphates, sulphates, potassium, sodium, calcium, magnesium, uric acid, ammonia, etc. The amount of urine voided is increased with some diseases and diminished by others. Similarly the colour of the urine varies according to its chemical composition—urates cause a reddish-yellow appearance, a greenish hue is due to the presence of bile and when blood is present the colour may be pink or red.

Healthy urine will leave a slight deposit or stain in any vessel in which it has been allowed to stand, due to the presence of urates, which sometimes become deposited in the urinary passages in the form of gravel.

Other substances such as albumin may be passed in the urine and these are discoverable by simple analytical tests. Their detection is important and early medical advice should be sought whenever unusual deposits make their appearance.

Suppression of the urine is the state in which the kidneys fail to act and retention describes the condition when the urine is retained in the bladder. Retention may be due to obstruction, pressure, nerve weakness, etc., and there should be no delay in calling in the doctor. Prostrate gland enlargement can cause blockage of the urethra, a condition common in elderly men.

Inability to retain the urine is sometimes due to muscular weakness or nerves and in simple cases the use of the appropriate tissue-salts is helpful, but an examination should always be made first for the presence of any other causative factor. *Nat. Sulph.* is the most frequently needed tissue-salt in the treatment of conditions affecting the urinary system.

Medical advice should be sought.

BIOCHEMIC TREATMENT

Ferr. Phos.   Incontinence of urine from muscular weakness.

Kali Phos.   Incontinence of urine from nervous debility.

Mag. Phos.   Constant urging to urinate when standing or walking. Spasmodic retention of urine.

Nat. Phos.   Incontinence of urine in children with acidity. Catarrh of the bladder.

Nat. Sulph.   Sandy deposit in the urine, gravel. Excessive excretions of urine.

Calc. Phos. Enuresis in old people, frequent urging to urinate.

Nat. Mur. Incontinence of urine in children, in alternation with *Kali Phos.* Involuntary emission while walking. Excessive flow of watery urine.

VERTIGO

Vertigo, or giddiness, may be due to various causes. The ability to balance depends upon sensations derived through the eyes, from touching, but mainly from the semicircular canals of the internal ears. Giddiness may be due to stomach upsets, nausea, headache, etc. Eyesight is a factor and disorders of the circulation may cause a temporary state of bloodlessness of the brain. Getting up suddenly from a sitting or bending position can cause slight giddiness, and elderly people who have to get out of bed during the night should pause momentarily in a sitting position to guard against a sudden faint. Smoking can also be a cause of giddiness. Medical advice should be sought.

BIOCHEMIC TREATMENT

Ferr. Phos. Giddiness from rush of blood to the head, with flushing, throbbing or pressing pain.

Kali Phos. Giddiness, swimming of the head, from nervous causes, worse when rising or looking upwards.

Nat. Sulph. Giddiness, with bitter taste in the mouth; gastric derangement with inclination to fall on the right side.

Mag. Phos. Vertigo from optical defects; dark spots floating before the eyes.

Nat. Phos. Giddiness with gastric derangements, acidity, loss of appetite. Creamy or golden coating of the tongue. Alternate with *Nat. Sulph.*

MINOR WOUNDS, SPRAINS and STRAINS

BIOCHEMIC TREATMENT

Ferr. Phos. is the first remedy for sprains, bruises, cuts, wounds, etc. It alleviates pain and congestion and should be applied, in powder form, externally wherever possible.

Kali Mur. For the swelling in alternation with *Ferr. Phos.*

Calc. Sulph. Bruises, cuts, wounds, etc., when neglected and suppurating.

Calc. Fluor. Bruises affecting the bones.

Silica. Neglected wounds with festerings. Discharges of thick yellow pus.

Nat. Sulph. For the shock and after-effects.

Calc. Phos. For fractures of the bones to help the fractures to mend. This tissue-salt should be given in all cases where there is injury or brittleness of the bones.

SYMPTOMATIC TREATMENT
OF MINOR AILMENTS

It is not possible, in a handbook of this size to give an exhaustive list of ailments, and their appropriate treatment with the tissue-salts. Nor is it essential that such a list be presented. In biochemic therapy, it is the prominent symptoms that are studied; these are the significant pointers to the remedy or remedies needed in any given case. It should be borne in mind that these symptoms are usually associated with some tissue-salt deficiency, and that they will cease to manifest as soon as this deficiency has been corrected.

Each of the tissue-salts has its own distinctive "symptom picture"—as for instance the spasms, cramps and neuralgias of *Mag Phos.*—and it is a knowledge of these "symptom pictures" which is the surest guide to the treatment needed. The student of biochemistry soon learns that the name given to an ailment is of little practical importance and he comes to think in terms of the individual tissue-salts and the particular sphere of action of each of them. Thus one may hear the experienced biochemic practitioner referring to a *"Ferr. Phos. subject"* or a *"Calc. Phos. subject"* or a *"Kali Phos. subject,"* and so on, according to the predominant deficiency revealed by the symptoms.

Moreover, to those who adopt the biochemic system of medicine, one thing soon becomes abundantly clear and it is this: If the early symptoms of a minor health disturbance are intelligently studied and appropriate corrective measures promptly applied, there is a very good chance that the "disease" will never happen.

The section headed "The Twelve Tissue-Salts. Their Place and Function in the Human Economy" should be read and re-read. Thus you will get to know how the remedies work and the predominant symptoms associated with each of them. Armed with this knowledge, you will soon learn to recognise when a particular tissue-salt is needed.

Many ailments have their beginning in some disturbance of the circulation, such as may be occasioned by a sudden chilling. If such a condition is not dealt with promptly, congestion and inflammation may follow and the way is then paved for local infection. Here is the "little acorn" from which the "massive oak" (chronic ill-health), may grow and the lesson to be learned from this simple fact is: *Never to neglect those so-called minor ailments.*

SPRING-CLEANING THE BLOOD STREAM

Much ill-health, particularly in adults, is merely a manifestation of the condition known as autotoxæmia or self-poisoning. From the moment of birth, food is taken into the alimentary canal with never a respite for the overworked organs of digestion. Debris begins to accumulate faster than it can be eliminated. Unwanted nutriment is stored away in reserve as fat, but sooner or later a radical cleansing effort has to be made by the body to rid itself of accumulations of waste products, and many so-called "diseases" are simply the outward and visible signs of this inward cleansing process. Unwanted organic matter, which incidentally provides a favourable soil for infection by micro-organisms, is removed in various ways and all the organs of elimination are used—bowels, kidneys, lungs and skin. Catarrh and eczema are examples of such an operation involving elimination through the mucous membranes and skin. No attempt should be made to *suppress* such symptoms. To do so only invites trouble of a more serious kind in the future. In fact, the suppression of symptoms is responsible for some of the degenerative

75

diseases in later life, e.g. rheumatism. Rather, we should assist the body in its periodical spring-cleaning and this can best be done by a period of rest and warmth and the lightest of diets. In fact, abstention from solid foods for a short period can make an important contribution.

PREVENTION IS BETTER THAN CURE

None will dispute the fact that prevention is better than cure but how few put this precept into practice. Have you the means at hand for dealing promptly with health emergencies? That touch of feverishness; lack of appetite; listlessness; headache; irritability and the many other daily ills with which every mother is all too familiar—these are Nature's distress signs and should never be ignored.

NOT A MOMENT TO LOSE

By the prompt use of the biochemic remedies, these early symptoms can generally be cleared up and the danger of serious developments averted. The chronic ailments of adult life can usually be traced back to those "childish ailments" and more particularly to those so-called "growing pains" which were all too often taken for granted and given scant attention.

WE KNOW BETTER TODAY

Today, fortunately for the young folk, we no longer speak of "growing pains" but recognise these pains for what they really are, the early signs of rheumatism and, in such cases, every mother, with a knowledge of the tissue-salts, will promptly give *Ferr. Phos.* This is the wise course, nip the trouble in the bud and so safeguard future health.

BETTER TO BE SURE THAN SORRY

There can be no doubt that a vast amount of human suffering would be avoided if the right kind of first-aid was at hand at the onset of illness. As a means of providing this medicinal first-aid the biochemic remedies are ideal. They are handy, pleasant to take and thoroughly reliable. And, remember, these remedies are not drugs, they are safe to use at all times and can have no ill-effects whatsoever. Actually, they are cell-foods; that is to say, they are chemical substances which have place and function in the life processes of the tissue-cells.

BE THE GOOD NEIGHBOUR

By carefully noting the action of the tissue-salts as you use them and passing on to others the results of your observations, you will be making your contribution to the sum of knowledge concerning this system of natural healing. Furthermore, you will be participating in the most noble of all causes—the relief of human suffering.

MAINTENANCE OF HEALTH IN MIDDLE AGE

Middle age is the period of life when we are all forced to realise that our reserves of strength are not inexhaustible. The snap and resilience of former years are no longer in evidence and it suddenly dawns upon us that we must do something to conserve our resources. There may be no serious break in health but there is a tendency to succumb more frequently to minor ailments and these are not so easily shaken off. Consequently, these minor ills must be given more thought and attention if they are not to have a cumulative, undermining effect, as a result of seriously disturbed metabolism.

77

Summed up in this phrase, we probably have the real cause of the ageing process; it is the most effective weapon of Old Father Time. What, precisely, is this metabolism, which is so apt to become disturbed with the passage of time? Briefly, it means the conversion of food into living tissue; it is a physiological process of analysis and synthesis. It embraces digestion, assimilation, tissue renovation and the provision of bodily heat and energy, in other words it is the sum total of the biochemical processes.

From this, it becomes obvious that if the human machine is to be kept running smoothly, disturbances of metabolism must be avoided and this calls for the correction of tissue-salt deficiencies with as little delay as possible.

It is a simple fact that, in middle age, the tendency to tissue-salt deficiencies is increased. As the years pass, many of these tissue-salts may become seriously depleted, hence the failing strength and signs of infirmity which are apt to make their appearance. In view of this the increasing frequency of signs of not-so-good health should be regarded as a warning. Those off-days can be a blessing in disguise if they are not taken for granted. That spell of depression, that disturbed stomach, those muscular twinges, all tell the same story—the body is calling for help. Heed these signs and take immediate steps to correct the tissue-salt deficiencies they reveal.

In what follows, some of the more frequently occurring disabilities of the not-so-young are considered and suggestions are made as to their biochemical treatment. There can be no doubt that the tissue-salts, used promptly, as occasion arises, can be the means of maintaining a reasonable standard of health in middle life, of keeping the faculties alert and of giving an evergreen touch to the "joy of living".

COLDS

If you are subject to frequent colds, coughs or other respiratory troubles—a short course of *Ferr. Phos.* and *Kali Mur.* is indicated.

RUNDOWN CONDITIONS

If you are rundown, weak or anæmic—*Ferr. Phos.* and *Calc. Phos.* will help to increase nutritional tone.

HEADACHES

If you are subject to headaches, neuralgia or neuritis —the tissue-salts *Kali Phos.* and *Mag. Phos.* will prove helpful.

STOMACH TROUBLES

If you are subject to indigestion or loss of appetite— a course of *Nat. Phos.* and *Ferr. Phos.* is indicated.

NERVOUS CONDITIONS

If you are nervous or irritable, suffer from mental depression or sleeplessness—take daily the tissue-salts *Kali Phos.* and *Mag. Phos.*

LIVER TROUBLES

If you are a "liver" subject, a course of *Nat. Sulph.*, the liver salt, and *Kali Mur.* is needed.

GIDDINESS

If you are subject to attacks of giddiness or dizziness, a course of *Ferr. Phos.* and *Kali Phos.* should prove helpful.

DOMESTIC ANIMALS

Many a pet lover has had cause to be thankful to the tissue-salts for their remedial aid in treating sick domestic animals. Their ailments are similar to our own, and their symptoms, viewed biochemically, are a guide to the treatment required. Animals respond well to such corrective measures. They take to this form of nutritional medication and their natural mode of living favours a quick response. The common minor ailments of domestic animals can be treated biochemically with most gratifying results but in the event of serious disease, or if the symptoms are in any way unusual, you should immediately consult a veterinary practitioner.

The biochemic tissue-salts are not drugs but vital cell-foods and their action is in complete harmony with *Nature*. The tiny tablets dissolve on the tongue, or may be given dissolved in a little warm water. For chronic cases, give three doses daily; acute cases, every half hour.

BIOCHEMIC TREATMENT

1. *Calc. Fluor.* Diseases affecting the surface of bones and enamel of teeth. Piles, hard glands, uterine displacements, prolapsus.

2. *Calc. Phos.* General tonic. Anæmia, rickets, broken bones, malnutrition. Coldness of the body. *Strengthens the teeth.*

3. *Calc. Sulph.* Mouth ulcers. Suppurations and wounds that are slow to heal. Abscess about the anus.

4. *Ferr. Phos.* Fevers, high temperatures, cuts, scratches, bleeding, (apply externally as powder or lotion). *The pre-eminent first-aid.*

5. *Kali Mur.* Flu (alternate with No. 4), respiratory ailments, greyish patches or spots in throat, constipation (light coloured stools), abscess, boils, (alternate with No. 12) eczema (alternate with Nos. 7 and 9). Sluggish liver. White or greyish-white discharges.

6. *Kali Phos.* Symptoms of nervous character. Highly strung animals, hysteria, asthma, shortness of breath.

7. *Kali Sulph.* Skin troubles, eczema (alternate with Nos. 5 and 9). Sore paws with scaling of skin. Catarrh, yellow crusts on the eyelids with yellowish, sticky discharge. Dry nose.

8. *Mag. Phos.* Nerve pains, cramp, spasmodic, nervous twitchings. Flatulent colic. Trembling of limbs.

9. *Nat. Mur.* Constipation with dry stools. Nettlerash, bites and stings of insects (also apply externally). Eczema with watery symptoms (alternate with Nos. 5 and 7). Anaemia (alternate with No. 2). Blood thin and watery, excess saliva.

10. *Nat. Phos.* Acidity, digestive upsets, worms, associated with fretfulness and irritability. Sour-smelling stools. Skin irritations with acid perspiration.

11. *Nat. Sulph.* Liverishness, biliousness with vomiting of watery, greenish fluid. Yellow eyeballs, humid asthma, rheumatic symptoms, worse from damp.

12. *Silica.* Pustules, boils, styes, suppurating wounds, ulcers. Diseased claws. *Silica and Kali Sulph, are excellent for conditioning the coat.*

SOME THOUGHTS ON FOOD AND FEEDING

THERE IS probably more nonsense talked about dieting than is talked about any other subject of everyday conversation. Fortunately, however, this business of eating for health is not so complicated nor as dull and dreary as one might be led to believe. To those seeking guidance on this subject, we would say most emphatically—*beware of "fads"*.

A good mixed diet is necessary for the proper nourishment of the human body. Further, the saying "Variety is the spice of life" applies very particularly to foods. Never forget that eating has its psychological aspect. Avoid monotony, let your food be interesting and as varied as possible—enjoy it!

Unless it has been grossly abused, the palate is a sure guide to the body's needs and can best tell you what to take and what to avoid at any given time. That "mouth-watering" interest in some particular food is of real significance. Therefore, keep the palate unsullied, do not bludgeon it with highly seasoned tit-bits and so-called appetisers.

Foods, as nature provides them, contain essential accessory factors (vitamins and minerals), and, in the case of raw fruits, water in its purest form. It is wise, therefore, to give preference always to fresh wholesome foods in their natural state, or as near so as possible, including that very necessary material known as "roughage" as provided in wholemeal bread, raw fruits, salad greens, fresh vegetables, etc.

Remember, the teeth are the advance guard of the digestion, see that they do their duty; thorough mastication is the answer to many digestive troubles. Leisurely eating is more satisfying and encourages the very desirable habit of moderation.

Plain, homely cooking ("As mother makes it") is always to be preferred. Good, wholesome food, plainly cooked and served, provides its own digestives and brings joy to that so frequently overworked organ, the stomach. To sum up, keep your "tummy" happy, it is here that health is made or marred.

> *"Now good digestion wait on appetite,*
> *And health on both".*

—Macbeth. Act iii, Sc.4, L.38.

COMBINATION REMEDIES

MANY YEARS of clinical experience and observed results have led to the introduction of skillfully formulated combined remedies for use in certain groups of ailments. The convenience of the combined form is self-evident for people who carry their remedies with them and when more than one of the tissue-salts are needed. The combined remedies are designated alphabetically. All are formulated strictly in accordance with homeopathic practice and they may be taken with complete safety in conjunction with other forms of treatment or in alternation with any of the single tissue salts.

A INSOMNIA, neurasthenic conditions, mental depression
Ferrum Phos., Kali Phos., Magnesia Phos.,

B DEBILITY, anemic conditions, nervous exhaustion and during convalesence
Calcarea Phos., Kali Phos., Ferrum Phos.,

C ACIDITY, gastric disorders, flatulence, biliousness with headaches
Natrum Phos., Natrum Sulph., Silicea

D ACNE, scaling of the skin, etc. for minor skin ailments and scalp eruptions, eczema
Kali Mur., Kali Sulph., Calcarea Sulph., Silicea

E	TONIC of the five tissue-salt phosphates in conditions of lowered vitality Calcarea Phos., Magnesia Phos., Ferrum Phos., Natrum Phos., Kali Phos.,
G	ELASTIC conditions, for backache, lumbago etc. Calcarea Fluor., Calcarea Phos., Kali Phos., Natrum Mur.
J	COLDS, acute rhinitis, coryza, acute cough, Ferrum Phos., Kali Mur., Natrum Mur.,
K	THROAT and tonsillar irritation, swelling of tonsils, and white coated tongue Ferrum Phos., Kali Mur., Kali Phos.

REPERTORY OF SYMPTOMS
AND THEIR
CORRESPONDING REMEDIES

HEAD

Blind headache: *Ferr. Phos.*
Cold applications relieve: *Ferr. Phos.*
Crawling feeling over head, with cold sensations: *Calc. Phos.*
Crusts, yellow, on scalp: *Calc. Sulph.*
Dandruff: *Nat. Mur., Kali Sulph.*
Dizziness: *Kali Phos.*
Eruptions on scalp, with watery contents: *Nat. Mur.*
Eruption of nodules on the scalp, with falling out of the hair: *Silica.*
  ,, on the head, with secretions of decidedly yellow, thin matter: *Kali Sulph.*
Giddiness, with gastric derangements: *Nat. Phos.*
Hair, falling of: *Kali Phos., Silica.*
  ,, loss of: *Calc. Phos.*
  ,, pulling causes pain: *Ferr. Phos.*
Head, cold to touch: *Calc. Phos.*
  ,, inability to hold up: *Calc. Phos.*
  ,, sore to touch: *Ferr. Phos.*
  ,, sweat on, of children: *Calc. Phos., Silica.*
  ,, trembling of: *Mag. Phos., Kali Phos.*
  ,, ulcers on top of: *Calc. Phos.*
Headache accompanied by:
  ,, biliousness, bitter taste: *Nat. Sulph.*
  ,, chills up and down spine: *Mag. Phos.*
  ,, cold feeling on head: *Calc. Phos.*
  ,, confusion: *Kali Phos.*
  ,, constipation: *Nat. Mur., Kali Mur.*
  ,, dizziness: *Nat. Sulph.*

Headache accompanied by:

" drowsiness: *Nat. Mur.*
" dull, heavy hammering: *Nat. Mur., Ferr. Phos.*
" feeling as if skull were too full: *Nat. Phos.*
" frothy coating on tongue: *Nat. Mur.*
" hammering, throbbing: *Ferr. Phos.*
" inability for thought: *Kali Phos.*
" intermittent and spasmodic pains: *Mag. Phos.*
" irritability: *Kali Phos.*
" loss of strength: *Kali Phos., Calc. Phos.*
" nodules, on head: *Silica.*
" pain in temples: *Ferr. Phos., Nat. Phos.*
" " over eye: *Ferr. Phos.*
" " in stomach: *Nat. Phos.*
" " throbbing, beating: *Ferr. Phos.*
" " on top of head: *Ferr. Phos., Nat. Sulph.*
" profusion of tears: *Nat. Mur.*
" prostrated feeling: *Kali Phos.*
" rush of blood to head: *Ferr. Phos.*
" sharp, shooting pains: *Mag. Phos.*
" vomiting of acid sour fluids: *Nat. Phos.*
" undigested food: *Nat. Phos., Ferr. Phos.*

Headache with:

" tearful mood: *Kali Phos.*
" thick white coating on the tongue: *Kali Mur.*
" unrefreshing sleep: *Nat. Mur.*
" vomiting of frothy phlegm: *Nat. Mur.*
" weariness: *Kali Phos.*
" yawning and stretching: *Kali Phos.*

Headache:

" aggravated by mental work: *Calc. Phos., Kali Phos.*
" " in evening: *Kali Sulph.*
" " " heated rooms: *Kali Sulph.*
" from loss of sleep: *Kali Phos.*
" from mental work: *Kali Phos.*
" in nervous subjects: *Kali Phos.*
" neuralgic: *Kali Phos., Mag. Phos.*
" " with humming in the ears: *Kali Phos., Ferr. Phos.*

Headache:

" of girls at puberty: *Nat. Mur., Calc. Phos.*
" " nervous character, with illusions of light:
      Mag. Phos.
" on awakening in the morning: *Nat. Phos.*
" " crown of head: *Nat. Phos.*
" " top of head, with pressure: *Nat. Phos.*
" " " with heat: *Nat. Phos.*

Headache relieved by cheerful excitement: *Kali Phos.*
" " " cool air: *Kali Sulph.*
" rheumatic, evening aggravations: *Kali Sulph.*
" sick, from sluggish action of liver: *Kali Mur.*
" " with bitter taste in mouth: *Nat. Sulph.*

Heaviness of the head in the morning after waking, with
      giddiness and dullness: *Nat. Mur.*

Inflammatory condition of the scalp: *Ferr. Phos.*

Mouth, bitter taste in: *Nat. Sulph.*

Neck, sharp pain in nape of: *Mag. Phos.*

Neuralgia of head, when pain is sharp: *Mag. Phos.*

Neuralgic headache, with humming in the ears, better
      under cheerful excitement, worse alone,
      tearful mood: *Kali Phos.*

Noises in head when falling asleep: *Kali Phos.*

Pain in the nape of the neck of a sharp character:
      Mag. Phos.

Pain and weight in the back part of the head, with
      weariness and exhaustion: *Kali Phos.*

Pain, aggravated by cold: *Mag. Phos.*
" relieved by cheerful excitement: *Kali Phos.*
" " " gentle motion: *Kali Phos.*
" " " heat: *Mag. Phos.*

Scalp, eruption on: *Silica.*
" inflammatory conditions of: *Ferr. Phos.*
" nodules on: *Silica.*
" painful, pustules on: *Silica.*
" sensitive: *Silica.*
" sore to touch: *Silica, Ferr. Phos.*
" suppurations of, discharge yellow and purulent:
      Calc. Sulph.

Scalp, tight sensations of: *Calc. Phos.*

„ white scales on: *Nat. Mur., Kali Mur., Kali Sulph.*

Sick headache arising from sluggish action of the liver, want of bile frequently accompanied by constipation: *Kali Mur.*

„ when the material vomited is undigested food: *Ferr. Phos.*

„ with bitter taste in the mouth, vomiting of bile or bilious diarrhœa: *Nat. Sulph.*

„ „ vomiting of sour fluids: *Nat. Phos.*

Skull, thin and soft: *Calc. Phos.*

Sleeplessness: *Kali Phos.*

Stitches in the head: *Nat. Mur.*

Trembling and involuntary shaking of the head: *Mag. Phos.*

Vertigo: *Calc. Phos.*

„ giddiness from excessive secretions of bile, tongue has a dirty greenish or greenish-brown coating at the back part, bitter taste in the mouth: *Nat. Sulph.*

„ from exhaustion and weakness: *Kali Phos.*

Violent pains at the base of the brain: *Nat. Sulph.*

MENTAL STATES

Anxious about future: *Calc. Phos.*

Backwardness: *Kali Phos.*

Brain-fag, from overwork: *Kali Phos.*

Children, crossness of: *Kali Phos.*

„ crying and screaming: *Kali Phos.*

„ ill-tempered: *Kali Phos.*

„ peevish and fretful: *Calc. Phos.*

„ screaming of, at night, during sleep: *Kali Phos., Nat. Phos.*

„ somnambulism in: *Kali Phos.*

Depressed spirits: *Kali Phos., Calc Phos., Nat. Mur.*

Desires solitude: *Calc. Phos.*

Despondent moods: *Nat. Mur., Nat. Sulph., Silica*

Discouraged, feels: *Nat. Sulph.*

Dizziness: *Ferr. Phos., Kali Phos.*
Fainting of nervous sensitive persons: *Kali Phos.*
   ,,   tendency to: *Kali Phos.*
Fits of crying: *Kali Phos.*
   ,, ,, laughing: *Kali Phos.*
Grasping at imaginary objects: *Kali Phos.*
Home-sickness: *Kali Phos.*
Hopeless, with dejected spirits: *Nat. Mur.*
Illusions, mental: *Mag. Phos., Kali Phos.*
Impatience and nervousness: *Kali Phos.*
Irritability: *Kali Phos., Nat. Phos.*
Irritation, due to biliousness: *Nat. Sulph.*
Melancholy: *Nat. Mur., Kali Phos.*
Memory, poor: *Calc. Phos., Kali Phos., Mag. Phos.*
Mind, wanders from one subject to another: *Calc. Phos.*
Moods, anxious: *Kali Phos.*
   ,, gloomy: *Kali Phos.*
Overstrain, from mental employment: *Kali Phos.*
Sensitiveness: *Kali Phos.*
Shyness: *Kali Phos.*
Sleeplessness: *Kali Phos.*
Sleepiness: *Nat. Mur.*
Stupor: *Nat. Mur.*
Thought, cannot concentrate: *Calc. Phos.*
   ,,   difficulty of: *Silica.*
Weeps easily: *Nat. Mur.*

EYES

Acrid tears in the eyes: *Nat. Mur.*
Acute pain in eyes: *Ferr. Phos.*
Agglutination at night with smarting of the lids: *Silica.*
Agglutination of lids in morning: *Nat. Phos.*
Black spots before eyes: *Kali Phos.*
Blisters on the cornea: *Nat. Mur.*
Bloodshot: *Ferr. Phos.*
Blurred vision, after straining eye: *Calc. Fluor.*
Burning of edges of lids: *Nat. Sulph.*
Colours before eyes: *Mag. Phos.*

Contracted pupils: *Mag. Phos.*
Cornea, blisters on: *Nat. Mur.*
    „ crusts on eyelids, yellow: *Kali Sulph.*
    „ inflammation of, with thick yellow discharges: *Calc. Sulph.*
Dimness of sight from weakness of the optic nerve: *Kali Phos.*
Discharge, golden-yellow, creamy: *Nat. Phos.*
    „ thick white mucus: *Kali Mur.*
    „ „ yellow: *Calc. Sulph.*
    „ greenish, serous: *Kali Sulph.*
    „ slimy secretions: *Kali Sulph.*
Double vision: *Mag. Phos.*
Drooping of lids: *Kali Phos.*, *Mag. Phos.*
Dry inflammation of eyes: *Ferr. Phos.*, *Nat. Mur.*
Excited appearance of eye: *Kali Phos.*
Eye affections, with flow of tears: *Nat. Mur.*
Eyes, bloodshot: *Ferr. Phos.*
    „ glued together in the morning, with a creamy discharge: *Nat. Phos.*
Eye-balls, ache: *Calc. Phos.*
    „ pain in the, relieved by resting eyes: *Calc. Fluor.*
Eyelids, specks of matter on: *Kali Mur.*
    „ yellow, mattery scabs on: *Kali Mur.*
Flow of tears from the eyes when associated with colds in the head: *Nat. Mur.*
    „ from weakness: *Nat. Mur.*
    „ on going into open air: *Nat. Mur.*
    „ with fresh colds: *Nat. Mur.*
    „ with neuralgic pains in eye: *Nat. Mur.*, *Mag. Phos.*
Granulations on eyelids: *Ferr. Phos.*, *Kali Mur.*
Inflammation of the eye, acute, with great intolerance of light: *Ferr. Phos.*
Inflammation of the eyes, when pus is discharging: *Calc. Sulph.*
    „ secreting of a golden-yellow, creamy matter: *Nat. Phos.*
    „ with discharge of thick yellow matter: *Silica.*

Lids, hot feeling of: *Calc. Phos.*
Light, great intolerance of: *Ferr. Phos.*, *Calc. Phos.*
„ sensitive to artificial: *Calc. Phos.*, *Mag. Phos.*
Neuralgic pains in eyes: *Calc. Phos.*, *Mag. Phos.*
Neuralgic pain in the eyes, with flow of tears: *Nat. Mur.*
Optic nerve, dullness of sight, from weakness of: *Kali Phos.*, *Mag. Phos.*
Pain as from excoriation in the eyes: *Nat. Mur.*
„ in the eyes, with tears: recurring daily at certain times: *Nat. Mur.*
Pupils, contracted: *Mag. Phos.*
„ dilated during disease: *Kali Phos.*
Redness and inflammation of the whites of the eyes with sensation as if the eye-balls were too large: *Nat. Mur.*
Sensitive to artificial light: *Calc. Phos.*, *Mag. Phos.*
Smarting secretions, with tears: *Nat. Mur.*
Sparks before eyes: *Mag. Phos.*
Sore eyes, with specks of matter on the lids or yellow mattery scabs: *Kali Mur.*
Spasms of eyelids: *Calc. Phos.*, *Mag. Phos.*
Spasmodic twitching of lids: *Mag. Phos.*, *Calc. Phos.*
Stoppage of tear ducts from cold: *Nat. Mur.*
Squinting: *Calc. Phos.*, *Mag. Phos.*
„ caused by irritation, from worms: *Nat. Phos.*
Staring appearance of eyes: *Kali Phos.*
Stye on lids: *Silica.*
Weak eyes, with tears when going into the cold air, or when wind strikes the eyes: *Nat. Mur.*
Yellow crusts on the eyelids: *Kali Sulph.*
Yellow-green matter in the eye: *Kali Sulph.*

EARS

Beating in the ears: *Silica.*
Boils around external ear: *Silica.*
Catarrh of ear, causing deafness: *Kali Sulph.*
„ involving eustachian tubes: *Kali Sulph.*

Catarrh involving middle ear: *Ferr. Phos., Kali Mur.*
Cracking noises in ear on blowing nose: *Kali Mur.*
     „   when swallowing: *Kali Mur.*
Cutting pain under ears: *Kali Sulph.*
Difficulty of hearing, accompanied by exhaustion of
         nervous system: *Kali Phos.*
     „   accompanied by thick, yellow discharge: *Calc.*
         Sulph.
     „   from inflammatory action: *Ferr. Phos.*
     „   „   swelling of eustachian tubes: *Nat. Mur., Kali*
         Mur., Silica, Kali Sulph.
Discharges, foul, ichorous, offensive: *Kali Phos.*
     „   mixed with blood: *Kali Phos.*
     „   thick, yellow, bloody: *Calc. Sulph.*
Ears, swollen, burning, itching: *Calc. Phos.*
Earache, accompanied by albuminous discharge: *Calc.*
         Phos.
     „   beating, throbbing pain: *Ferr. Phos.*
Earache, accompanied by excoriating discharge: *Calc.*
         Phos.
Earache, grey or white-furred tongue: *Kali Mur.*
     „   lightning-like pain through ears: *Nat. Sulph.,*
         Mag. Phos.
Earache, accompanied by swelling of eustachian tube;
         glands or tonsils: *Kali Mur.*
     „   yellow, mattery discharge: *Kali Sulph.*
     „   aggravated by cold: *Mag. Phos.*
     „   „   „   damp weather: *Nat. Sulph.*
     „   of nervous or spasmodic character: *Mag. Phos.*
     „   relieved by heat: *Mag. Phos.*
Exudations from ear, thick, white, and moist: *Kali Mur.*
Glands around the ear swollen; noises in the ear;
         snapping and cracking: *Kali Mur.*
Granulations moist, grey or thick white exudation from
         the ear: *Kali Mur.*
Heat and burning of the ears, with gastric symptoms:
     Nat. Phos.
Humming in the ears: *Nat. Mur.*

Inflammation of the ears, first stage for the fever and pain: *Ferr. Phos.*

„ external ear with redness and burning: *Ferr. Phos.*

„ loud noise aggravates: *Silica.*

Noises in ears and head, with confusion: *Kali Phos.*

„ like running water: *Ferr. Phos.*

Outer ear sore and scabby: *Nat. Phos.*

„ with creamy discharge: *Nat. Phos.*

Scabs, with creamy, yellow appearance: *Nat. Phos.*

Sharp pain under ears: *Kali Sulph.*

Singing or tingling in the ears: *Nat. Mur.*

Stitches in the ears: *Nat. Mur.*

Swelling of the parotid gland, with stitching pain: *Silica.*

Ulceration of the ear, when the discharge is foul, ichorous, offensive, sanious, or mixed with blood: *Kali Phos.*

Whizzing and ringing in the ears with diminution of hearing: *Mag. Phos.*

NOSE

Bleeding from the nose: *Ferr. Phos.*

„ in delicate constitutions, when the blood is thin, blackish or coagulating; predisposition to bleeding: *Kali Phos.*

Bleeding in anæmic persons, the blood is thin and watery: *Nat. Mur.*

„ during menses: *Nat. Sulph.*

Boils on edges of nostrils: *Silica.*

Burning in nose: *Nat. Sulph.*

Catarrh, accompanied by fever: *Ferr. Phos.*

„ acute or chronic, with slimy yellow, greenish discharges: *Ferr. Phos., Kali Sulph.*

„ albuminous discharge, thick and tough, dropping from the posterior nares and causing constant hawking and spitting, worse out of doors: *Calc. Phos.*

Catarrh, aggravated in evening: *Kali Sulph.*
   „   „   „   warm room: *Kali Sulph.*
   „   dry, with stuffy sensation: *Kali Mur.*
   „   chronic, with purulent discharges from anterior or
                 posterior nares: *Kali Sulph., Silica.*
   „   of anæmic persons: *Nat. Mur., Calc. Phos.*
   „   with fetid discharges: *Kali Phos.*
   „   with salty, watery mucus: *Nat. Mur.*
   „   „   stuffy sensation: *Kali Mur.*
   „   „   white, not transparent phlegm: *Kali Mur.*
Cold in the head, with yellow creamy discharge from
            the nose; itching of the nose: *Nat. Phos.*
   „   in the third stage of resolution, when the discharge
            is thick, yellow, purulent, and sometimes
            tinged with blood: *Calc. Sulph.*
   „   with dry, harsh skin; to produce perspiration: *Kali*
            Sulph.
Crusts in the vault of the pharynx: *Kali Mur.*
Discharge, albuminous: *Calc. Phos.*
   „   clear, watery, transparent mucus: *Nat. Mur.*
   „   fetid: *Kali Phos.*
   „   slimy, yellow, watery, greenish: *Kali Sulph.*
   „   thick and white: *Kali Mur.*
   „   yellow, fetid: *Silica.*
   „   lumpy, green: *Calc. Fluor.*
   „   purulent, bloody: *Calc. Sulph.*
   „   yellow, creamy: *Nat. Phos.*
Disposition to catch cold in anæmic persons: *Calc.*
            Phos., Nat. Mur.
Dryness and burning in the nose: *Nat. Sulph.*
Dryness of nose, with scabbing: *Nat. Mur., Silica.*
Edges of nostrils itch: *Silica.*
First or inflammatory stage of colds: *Ferr. Phos.*
Fluent coryza: *Nat. Mur.*
Frequent sneezing: *Silica, Nat. Mur.*
Fresh cold and discharge of clear, watery transparent
            mucus, and sneezing: *Nat. Mur.*
Hawking and spitting, constant: *Calc. Phos.*

Hay fever: *Nat. Mur.*

Itching or redness at tip of nose: *Silica.*

  ,,   the nose: *Nat. Phos.*

Loss of smell or perversion of the sense of smell, not
               connected with a cold: *Mag. Phos.*

  ,,   ,,   ,,   with dryness and rawness of the pharynx:
               Nat. Mur.

Nose, inflamed at edges of nostrils: *Silica.*

  ,,   swollen: *Calc. Phos.*

  ,,   cold: *Calc. Fluor.*

Pharynx, dryness and rawness of: *Nat. Mur.*

Picks at nose: *Nat. Phos.*

Polypi: *Calc. Phos.*

Sneezing: *Nat. Mur.*

Stuffy cold in head, with yellow, lumpy, green dis-
               charges: *Calc. Fluor.*

  ,,   with collection of greenish mucus: *Kali Sulph.,*
               Silica.

Takes cold easily: *Ferr. Phos., Calc. Phos.*

FACE

Acne: *Calc. Sulph.*

Anæmic face: *Calc. Phos.*

Beard, tender pimples under: *Calc. Sulph.*

Blotches on face, come and go suddenly: *Nat. Phos.*

Chaps of lips: *Calc. Fluor.*

Cheek swollen and painful: *Kali Mur.*

Creeping pains in face: *Calc. Phos.*

Dirty-looking face: *Calc. Phos.*

Eruption on the face from any cause, with discharge:
               Silica.

Face, bloated, without fever: *Nat. Phos.*

  ,,   flushed, cold sensation at nape of neck: *Ferr.*
               Phos.

Face, livid: *Kali Phos.*

  ,,   pale, sickly, sallow: *Kali Phos., Calc. Phos.*

  ,,   pallid and pale: *Ferr. Phos., Calc. Phos.*

  ,,   red, without fever: *Nat. Phos.*

Faceache, accompanied by:

      ,,   constipation: *Nat. Mur.*
      ,,   flow of tears: *Nat. Mur.*
      ,,   cutting pains: *Mag. Phos.*
      ,,   small lumps on face: *Silica.*
      ,,   from swelling of cheek: *Kali Mur.*

Feeling of coldness or numbness of face: *Calc. Phos.*

Feverish complexion: *Ferr. Phos.*

Frothy bubbles at edge of tongue: *Nat. Mur.*

Grinding pains in face: *Mag. Phos., Calc. Phos.*

Hard swelling on cheeks, with toothache: *Calc. Fluor.*

Inflammatory neuralgia of the face: *Ferr. Phos.*

Lightning-like pains in face: *Mag. Phos.*

Neuralgia, accompanied by flow of tears: *Nat. Mur.*

     ,,   ,,   ,,   shifting pains: *Mag. Phos., Kali Sulph.*
     ,,   ,,   ,,   shooting pains: *Mag. Phos.*
     ,,   ,,   ,,   spasmodic pains: *Mag. Phos.*
     ,,   aggravated by being in heated room: *Kali Sulph.*
     ,,   ,,   ,,   cold: *Mag. Phos.*
     ,,   in the evening: *Kali Sulph.*
     ,,   with exhaustion of nervous system: *Kali Phos.*
     ,,   relieved by being in cool air: *Kali Sulph.*
     ,,   relieved by hot applications: *Mag. Phos.*

Nodules on face: *Calc. Sulph.*

Pains and heat in face: *Ferr. Phos.*

     ,,   cold applications soothe: *Ferr. Phos.*

Pale face in children when teething is difficult: *Calc. Phos.*

Pallid face, from a lack of red blood corpuscles: *Ferr. Phos.*

Pimples on face, mattery: *Calc. Sulph.*

     ,,   at age of puberty: *Calc. Sulph., Calc. Phos.*

Skin cold and clammy: *Calc. Phos.*

Tearing pain in face: *Mag. Phos., Calc. Phos.*

White about mouth and nose: *Nat. Phos.*

Yellow, sallow, or jaundiced face due to biliousness: *Nat. Sulph.*

MOUTH

Acid taste in mouth: *Nat. Phos.*
Bad taste in mouth: *Nat. Sulph.*, *Kali Phos.*
  ,, in morning: *Calc. Phos.*
Bitter taste in mouth: *Nat. Sulph.*
Blisters like pimples on the tip of the tongue: *Calc. Phos.*
Clean tongue with an inflammatory condition: *Ferr. Phos.*
Coating on the tongue white and slimy: *Kali Mur.*
  ,, yellow, sometimes with whitish edge: *Kali Sulph.*
Constant hawking of slimy mucus: *Nat. Sulph.*
Constant spitting of frothy mucus: *Nat. Mur.*
Cracked lips: *Calc. Fluor.*
Creamy, golden-yellow exudation from tonsils and pharynx: *Nat. Phos.*
Creamy, yellow coating at back part of roof of mouth: *Nat. Phos.*
Dirty greenish-grey or greenish-brown coating on the root of the tongue with saliva: *Nat. Sulph.*
Dryness of the lower lips; skin pulls off in large flakes: *Kali Phos.*
  ,, ,, tongue in fevers, with watery discharge from the bowels: *Nat. Mur.*
Glands and gums swollen: *Kali Mur.*
  ,, swelling of, under tongue: *Nat. Mur.*
Gums hot, swollen, and inflamed: *Ferr. Phos.*
Hard swelling on jaw-bones: *Calc. Fluor.*
Hawking, constant, of foul, slimy mucus from trachea and stomach: *Nat. Sulph.*
Inflammation of salivary glands, when secreting excessive amount of saliva: *Nat. Mur.*
Mouth full of thick, greenish-white, tenacious slime: *Nat. Sulph.*
Rawness of mouth: *Kali Mur.*
Saliva, excess of: *Kali Mur.*
Sour taste in mouth: *Nat. Phos.*
Spasms of stammering: *Mag. Phos.*
Speaks slowly: *Mag. Phos.*

Swelling of glands under the tongue: *Nat. Mur.*
Thrush in children: *Kali Mur.*
   ,, with much saliva: *Nat. Mur.*
Twitching, spasmodic, of lips: *Mag. Phos.*
   ,, mouth: *Mag. Phos.*
Ulcers in mouth, ashy-grey: *Kali Phos.*
   ,,   ,,   ,, white: *Kali Mur.*
   ,,   ,, corners of mouth: *Silica.*
Very offensive breath: *Kali Phos.*

TONGUE

Blisters on tip of tongue: *Nat. Mur., Calc. Phos.*
Chronic swelling of: *Calc. Fluor.*
Clean and red: *Ferr. Phos.*
Coating on tongue, clear, slimy, watery: *Nat. Mur.*
   ,, dirty, greenish-grey, bitter taste: *Nat. Sulph.*
   ,, golden-yellow, on back part: *Nat. Phos.*
   ,, greyish-white: *Kali Mur.*
   ,, like stale brownish liquid mustard, offensive breath: *Kali Phos.*
   ,, moist, creamy on back part: *Nat. Phos.*
   ,, yellow and slimy: *Kali Sulph.*
Cracked appearance of tongue: *Calc. Fluor.*
Dark red and inflamed: *Ferr. Phos.*
Dry in the morning: *Kali Phos.*
Dryish or slimy: *Kali Mur.*
Frothy bubbles on edges of: *Nat. Mur.*
Induration of tongue, after inflammations: *Silica, Calc. Fluor.*
Inflammation, for swelling: *Ferr. Phos., Kali Mur.*
   ,, with exhaustion: *Kali Phos.*
   ,, when suppurating: *Silica, Calc. Sulph.*
Numbness of tongue: *Calc. Phos.*
Pimples on tip of: *Calc. Phos.*
Stiffness of: *Calc. Phos.*
Swollen: *Kali Mur., Calc. Phos.*
Ulcers on: *Silica.*
Vesicles on tongue: *Nat. Mur.*

TEETH AND GUMS

Children grind teeth during sleep: *Nat. Phos.*
Cramps during dentition: *Mag. Phos.*
Decay of teeth as soon as they appear: *Calc. Phos.*
Dentition retarded: *Calc. Phos.*
Enamel, brittle: *Calc. Fluor.*
  „ rough and thin: *Calc. Fluor.*
Gastric derangements during teething: *Nat. Phos.*
Gums bleed easily: *Kali Phos.*
  „ pale: *Calc. Phos.*
  „ predisposition to bleed: *Kali Phos.*
Gum-boil: *Silica.*
  „ before pus begins to form: *Kali Mur.*
Infants, teething of, with drooling: *Nat. Mur.*
Loose in sockets: *Calc. Fluor.*
Nervous chattering of teeth: *Kali Phos.*
Neuralgia of teeth: *Nat. Mur.*
Rapid decay of teeth: *Calc. Fluor., Calc. Phos.*
Seam, bright-red, on gums: *Kali Phos.*
Sockets, teeth loose in: *Calc. Fluor.*
Teeth sensitive to cold air: *Mag. Phos.*
  „ „ „ touch: *Mag. Phos., Calc. Fluor.*
Toothache accompanied by:
  „ deep-seated pain: *Silica.*
  „ excessive flow of saliva or of tears: *Nat. Mur.*
  „ neuralgia of face: *Mag. Phos.*
  „ sharp, shooting pains, spasmodic: *Mag. Phos.*
  „ swelling of gums or cheeks: *Kali Mur., Ferr.
        Phos.*
  „ ulceration: *Silica.*
Toothache aggravated by being in warm room: *Kali
        Sulph.*
  „ „ „ hot liquids: *Ferr. Phos.*
  „ in nervous subjects: *Mag. Phos., Kali Phos.*
  „ relieved by being in open air: *Kali Sulph.*
  „ „ „ cold applications: *Ferr. Phos.*
  „ „ „ liquids: *Ferr. Phos.*
  „ „ „ hot applications: *Mag. Phos.*
Ulceration of roots of teeth: *Calc. Sulph.*
  „ with swelling gums and cheeks: *Calc. Sulph.*

THROAT

Burning sensation in the pharynx and cases of chronic catarrh, when there is considerable dropping from the posterior nares: *Calc. Phos.*

Choking on attempting to swallow: *Mag. Phos.*

Constricted feeling of throat: *Mag. Phos.*

Closing of larynx by spasms or cramp: *Mag. Phos.*

Constant hoarseness: *Calc. Phos.*

Dry red and inflamed throat: *Ferr. Phos.*

First stage of sore throat, when there is pain, heat, and redness: *Ferr. Phos.*

Glands painful, aching: *Calc. Phos.*

Hoarseness, constant: *Calc. Phos., Ferr. Phos.*

Inflammation of the mucous lining of the throat, with watery secretions: *Nat. Mur.*

     "   " tonsils: *Ferr. Phos.*

     "   " with swelling and greyish-white patches: *Kali Mur.*

Larynx, burning and soreness in: *Calc. Phos., Ferr. Phos.*

     " closing of, by spasm: *Mag. Phos.*

Loss of voice: *Kali Mur.*

     " from strain: *Ferr. Phos.*

Lump in, on swallowing: *Nat. Sulph.*

Pharynx, burning and soreness in: *Calc. Phos.*

Raw feeling in throat: *Nat. Phos.*

Redness and inflammation: *Ferr. Phos.*

Relaxed condition of: *Calc. Fluor.*

Scraping of, when talking: *Calc. Phos.*

Sticking pain in, on swallowing: *Calc. Phos.*

Shrill voice, coming on suddenly while speaking: *Mag. Phos., Kali Phos.*

Sore, raw feeling in the throat; tonsils and throat inflamed, with creamy, yellow, moist coating: *Nat. Phos.*

     " throat as if a plug had lodged in the throat: *Nat. Mur.*

     "   " of singers and speakers: *Ferr. Phos.*

Sore throat with excessive dryness or too much secretion: *Nat. Mur.*

Spasms of the throat: *Mag. Phos.*

Spasmodic cough: *Mag. Phos.*

Stinging sore throat, only when swallowing, the neck being painful to touch: *Silica.*

Suppuration of throat: *Calc. Sulph.*

Swallowing, painful: *Ferr. Phos.*

Thirst, with dry mouth: *Calc. Phos.*

Tonsillitis, after pus has begun to form: *Silica.*

Tonsils, chronic enlargement of: *Calc. Phos.*

   ,, creamy, yellow, moist coating on: *Nat. Phos.*

   ,, grey-white patches on: *Kali Mur.*

Tonsils, inflamed: *Nat. Phos., Ferr. Phos.*

Ulcerations, with thick yellow discharges: *Silica.*

Ulcerated throat, with fever and pain: *Ferr. Phos.*

   ,, white or grey patches: *Kali Mur.*

Windpipe, spasmodic closing of: *Mag. Phos.*

GASTRIC SYMPTOMS

Abnormal appetite, but food causes distress: *Calc. Phos.*

Acid drinks aggravate: *Mag. Phos.*

All conditions when excess of saliva and watery vomiting present; tongue has a clear, frothy, transparent coating: *Nat. Mur.*

   ,, ,, ,, of the stomach when there are sour acid risings or the tongue has a moist, creamy yellow coating: *Nat. Phos.*

Appetite not satisfied: *Kali Phos.*

Belching brings back taste of food: *Ferr. Phos.*

   ,, sour eructation: *Nat. Phos.*

Bilious colic: *Nat. Sulph.*

Biliousness from too much bile: *Nat. Sulph.*

Bitter taste in mouth: *Nat. Sulph.*

Bloated, stomach feels: *Calc. Phos.*

Burning in stomach: *Calc. Phos., Kali Sulph., Ferr. Phos.*

Catarrh of the stomach, with yellow, slimy tongue: *Kali Sulph.*

Clear, frothy, transparent coating on tongue: *Nat. Mur.*

Cold drinks relieve symptoms: *Ferr. Phos.*

   „   „   aggravate symptoms: *Calc. Phos., Mag. Phos.*

Constipation, with waterbrash: *Nat. Mur.*

Craving for salt or salty food: *Nat. Mur.*

Distress about heart: *Kali Phos.*

Dizziness: *Nat. Sulph.*

Dread of hot drinks: *Kali Sulph.*

Dyspepsia with:

   „ acid risings: *Nat. Phos.*

   „ flushed face and throbbing pain in the stomach: *Ferr. Phos.*

   „ pain after eating, if watery symptoms are present: *Nat. Mur.*

Dyspepsia with white-grey coating on the tongue, heavy pain under the right shoulder blade, eyes look large and protruding: *Kali Mur.*

Evacuations, bilious, green: *Nat. Sulph.*

Excess of saliva: *Nat. Mur.*

Faint, sick feeling in the region of the stomach: *Calc. Phos.*

Fatty food disagrees: *Kali Mur., Nat. Phos.*

Flatulence, with distress about heart: *Kali Phos., Nat. Phos.*

   „ with sluggishness of the liver: *Kali Mur., Nat. Sulph.*

Food aggravates: *Calc. Phos.*

   „ causes pain: *Nat. Phos.*

   „ distresses: *Calc. Phos.*

   „ vomiting of: *Ferr. Phos.*

Fullness at pit of stomach: *Kali Sulph.*

Gastric abrasions, superficial ulcerations if acid symptoms are present: *Nat. Phos.*

Great thirst: *Nat. Mur.*

Heartburn: *Silica, Ferr. Phos., Calc. Phos, Nat. Phos.*

Heaviness in stomach: *Calc. Phos.*

Hiccough: *Mag. Phos.*

Hungry feeling after eating: *Kali Phos.*

Indigestion, accompanied by griping pains: *Mag. Phos.*

    „ with pain in the stomach and watery gathering in the mouth, or sour taste in the mouth: *Nat. Mur.*

    „ with pressure and fullness at the pit of the stomach: *Kali Sulph.*

    „ vomiting of greasy, white, opaque mucus: *Kali Mur.*

    „ watery vomiting and salty taste in the mouth: *Nat. Mur.*

Infants vomit sour curdled milk: *Calc. Phos.*, *Nat. Phos.*

Liver, cutting pain in region of: *Nat. Sulph.*

Lump, food lies in a: *Calc. Phos.*

Milk, infants vomit curdled: *Calc. Phos.*, *Nat. Phos.*

Moist, creamy, yellow coating on tongue: *Nat. Phos.*

Morning sickness: *Nat. Phos.*

Mouth full of slimy mucus: *Nat. Sulph.*

Nausea, with sour risings: *Nat. Phos.*

    „ immediately after a meal: *Nat. Mur.*

Nausea, with "gone" sensation in the stomach: *Kali Phos.*

Neuralgia of stomach: *Mag. Phos.*

Nurse, constant desire of infants to: *Calc. Phos.*

Nurses, child vomits as soon as it: *Calc. Phos.*, *Ferr. Phos.*, *Silica.*

Pain in stomach after eating: *Nat. Mur.*, *Calc. Phos.*, *Nat. Phos.*

    „ is remittent and spasmodic: *Mag. Phos.*

    „ sometimes relieved by belching: *Calc. Phos.*

    „ worse from eating even the smallest amount of food: *Calc. Phos.*

Pastry disagrees: *Kali Mur.*

Pressure at pit of stomach: *Kali Sulph.*

Right shoulder-blade, pain under: *Kali Mur.*

Salty taste in mouth: *Nat. Mur.*

Sea sickness: *Kali Phos.*, *Nat. Sulph.*

Sick headache from gastric derangements: *Nat. Sulph.*
Sour, acid risings: *Nat. Phos.*
Spasms of stomach, with griping: *Mag. Phos.*
Stomach sore to touch: *Calc. Phos.*
    „ tender to touch: *Ferr. Phos.*
Stomach-ache accompanied by:
    „ „ „ constipation: *Kali Mur.*
    „ „ „ depression: *Kali Phos.*
    „ „ „ exhaustion: *Kali Phos.*
    „ „ „ loose evacuations: *Ferr. Phos.*
    „ from acidity of the stomach: *Nat. Phos.*
    „ „ chill: *Ferr. Phos.*
    „ „ worms: *Nat. Phos.*
Thirst, great: *Nat. Mur.*
Thirstlessness: *Kali Phos.*
Vomiting, after cold drinks: *Calc. Phos.*
    „ bile: *Nat. Sulph.*
    „ bright-red blood: *Ferr. Phos.*
    „ dark, clotted blood: *Kali Mur., Ferr. Phos.*
    „ fluids like coffee-grounds: *Nat. Phos.*
    „ from stomach-ache: *Mag. Phos.*
    „ greenish water: *Nat. Sulph.*
    „ sour acid fluids: *Nat. Phos.*
    „ thick white phlegm: *Kali Mur.*
Vomiting undigested food: *Ferr. Phos., Calc. Fluor.*
    „ watery: *Nat. Mur.*
Waterbrash, with constipation: *Nat. Mur.*
Water gathers in mouth: *Nat. Mur.*

ABDOMEN

Abdomen, bloated: *Kali Sulph., Mag. Phos.*
    „ cold to touch: *Kali Sulph.*
    „ cutting pains in: *Nat. Sulph., Mag. Phos., Ferr. Phos.*
    „ distended: *Mag. Phos.*
Abdomen, inflammation, fever: *Ferr. Phos.*
    „ sunken: *Calc. Phos.*

Abdomen, swollen: *Kali Phos., Kali Mur.*
  „ tender to touch: *Kali Mur.*
Anus, itching at: *Nat. Phos., Calc. Fluor.*
  „ cracks and fissures of the: *Calc. Fluor.*
  „ pain in: *Kali Mur.*
Back, pain in: *Calc. Fluor.*
Bilious evacuations: *Nat. Sulph.*
Bowels, loose in old people: *Nat. Sulph.*
  „ sore and tender: *Ferr. Phos.*
Burning in the bowels: *Silica.*
  „ sore pain in the pit of the stomach: *Ferr. Phos.*
Burning pain in rectum: *Nat. Mur.*
Colic of infants: *Mag. Phos.*
Constant urging to stool: *Kali Mur.*
Constipation, see Stools.
Diarrhœa, see Stools.
Distended abdomen: *Mag. Phos.*
Fæces, inability to expel: *Calc. Fluor.*
Flatulence, with pains in left side: *Kali Phos.*
Flatulent colic: *Nat. Phos., Nat. Sulph.*
Flatulent distension of the abdomen: *Nat. Mur.*
Frequent calls to stool, no passage: *Calc. Phos., Kali Phos., Mag. Phos.*
Heat in lower bowels: *Nat. Sulph., Ferr. Phos.*
Liver, pains in regions of: *Kali Mur.*
  „ sensitive: *Nat. Sulph.*
Liver, sharp, shooting pains in: *Nat. Sulph.*
  „ sluggish: *Kali Mur.*
  „ region of, sore to touch: *Nat. Sulph.*
Neuralgia of bowels: *Mag. Phos.*
  „ „ rectum: *Calc. Phos.*
Pain of a colicky nature, caused by sudden change from hot to cold: *Kali Sulph.*
Pains in abdomen relieved by pressure: *Mag. Phos.*
  „ „ rubbing: *Mag. Phos.*
  „ „ warmth: *Mag. Phos.*
Piles, bleeding: *Ferr. Phos., Calc. Fluor.*
Rectum, pain in: *Mag. Phos.*

Sluggish action of the liver, with pale yellow evacuations; pain in region of liver or under the right shoulder-blade: *Kali Mur.*

Spasmodic pains: *Mag. Phos.*

Sulphurous odour of gas from bowels: *Kali Sulph.*

Swelling of abdomen: *Kali Mur.*

Torn feeling after stool: *Nat. Mur.*

STOOLS

Bowels discharging mattery substances: *Calc. Sulph.*

Constipation from dryness of the mucous membrane, with watery secretions in other parts: *Nat. Mur.*

„ light-coloured stool, showing want of bile; sluggish action of the liver: *Kali Mur.*

„ with drowsiness and watery symptoms from the eyes or mouth: *Nat. Mur.*

„ „ dull, heavy headache, profusion of tears or vomiting of frothy mucus: *Nat. Mur.*

Diarrhœa after eating greasy, fatty food: *Kali Mur.*

„ alternating with constipation: *Nat. Mur.*

„ especially of children, with green, sour-smelling stools caused by an acid condition: *Nat. Phos.*

„ in teething children; stools slimy, green, undigested, with colic: *Calc. Phos.*

„ like water: *Nat. Mur.*

„ of schoolgirls, accompanied by headache: *Calc. Phos.*

„ stools frothy, slimy, causing soreness and smarting: *Nat. Phos.*

„ when there is much straining at stool or constant urging to stool, with passing of jelly-like mucus indicating acidity: *Nat. Phos.*

„ with greenish, bilious stools or vomiting of bile: *Nat. Sulph.*

Diarrhœa with pale, yellow, clay-coloured stool, swelling
of the abdomen, slimy stools: *Kali Mur.*
„ „ putrid, foul evacuations, depression and exhaustion of the nerves: *Kali Phos.*
„ „ yellow, slimy, purulent matter: *Kali Sulph.*
Flatulent colic, with green sour-smelling stools, or
vomiting of curdled masses: *Nat. Phos.*
Frequent call for stool, but passes nothing: *Calc. Phos.*
Griping pain in the abdomen, with watery diarrhœa,
stools expelled with force: *Nat. Mur.*
Loose morning stool, worse in cold wet weather: *Nat.
Sulph.*
Looseness of the bowels in old people: *Nat. Sulph.*
„ with watery stools: *Nat. Mur.*
Offensive stools: *Calc. Phos., Kali Phos.*
Retention of stool: *Nat. Mur.*
Stool is hot, often noisy and offensive: *Calc. Phos.*
Stools are dry and often produce fissures in the rectum:
Nat. Mur.

URINARY SYMPTOMS

Bladder, inflammation of: *Kali Mur., Ferr. Phos., Calc.
Sulph.*
Brickdust sediment in urine: *Nat. Sulph.*
Burning after urinating: *Nat. Mur., Ferr. Phos.*
„ pain over kidneys: *Ferr. Phos.*
Constant urging to urinate, if not chronic: *Ferr. Phos.*
Cutting pains after urinating: *Nat. Mur.*
„ pains at neck of bladder: *Calc. Phos.*
Dark red urine, with rheumatism: *Nat. Phos.*
Desire to urinate, with scanty emission: *Silica.*
Enuresis of children (wetting of the bed): *Kali Phos.,
Nat. Phos., Ferr. Phos., Nat. Mur.*
„ if from worms: *Nat. Phos.*
Excessive flow of watery urine: *Nat. Mur., Ferr. Phos.*
Frequent urination, with inability to retain the urine,
and acidity: *Nat. Phos.*

Frequent passing of much water, sometimes scalding: *Kali Phos.*

Gravel in bilious persons: *Nat. Sulph.*
  „   pain while passing: *Nat. Sulph., Mag. Phos.*
  „   sediment in urine: *Nat. Sulph., Calc. Phos.*
  „   with gouty symptoms: *Nat. Sulph.*

Great thirst, with excessive flow of watery urine: *Nat. Mur.*

Highly-coloured urine: *Calc. Phos., Ferr. Phos.*
  „   „   „   with fever: *Nat. Phos., Ferr. Phos.*

Inability to retain urine, from nervous debility: *Kali Phos.*

Incontinence weakness of sphincter: *Ferr. Phos.*

Increase in quantity of urine: *Calc. Phos.*

Involuntary emission of urine while walking: *Nat. Mur.*

Sandy deposits in urine: *Nat. Sulph.*

Sediment clings to side of vessel: *Nat. Sulph.*

Sharp shooting pains at neck of bladder: *Calc. Phos.*

Smarting on urinating: *Ferr. Phos.*

Spasmodic retention of urine: *Mag. Phos.*

Spasms of bladder, with painful straining: *Mag. Phos., Ferr. Phos., Kali Phos.*

Urine frequently scalding: *Kali Phos.*
  „   dark coloured, when there is torpidity and inactivity of the liver: *Kali Mur.*

FEMALE ORGANS

Abdominal pains followed by leucorrhœa: *Mag. Phos.*

Aching in uterus: *Calc. Phos.*

Acid leucorrhœa, worse after menstruating: *Calc. Phos.*

Acrid pain during leucorrhœa, with yellowness of the face: *Nat. Mur.*

After confinement when the pelvic muscles are relaxed: *Calc. Fluor.*

Colic in nervous, lachrymose women: *Kali Phos., Mag. Phos.*

109

Congestion of the uterus, menstrual periods too frequent; lasting too long: *Kali Mur.*

Discharge, deep-red or blackish-red: *Kali Phos.*
   „   scalding, smarting: *Nat. Mur.*
   „   sickening: *Nat. Phos.*
   „   sour-smelling: *Nat. Phos.*
   „   thick, white, bland: *Kali Mur.*
   „   thin, with offensive odour: *Kali Phos.*

Dragging in groin: *Calc. Fluor.*
   „   „   small of back: *Calc. Fluor.*

Dryness of vagina: *Nat. Mur.*

Dysmenorrhœa: *Mag. Phos.*
   „   labour-like pains during: *Mag. Phos.*
   „   with congestion: *Ferr. Phos.*
   „   vomiting of undigested food: *Ferr. Phos.*

Increased menses: *Silica.*

Leucorrhœa, accompanied by:
   „   albuminous discharge: *Calc. Phos.*
   „   milky-white, non-irritating discharge: *Kali Mur.*
   „   rawness and itching of parts: *Nat. Phos.*
   „   scalding, acrid discharge: *Nat. Mur., Kali Phos.*
   „   slimy, greenish discharge: *Kali Sulph.*
   „   thick, yellow, bloody discharge: *Calc. Sulph.*
   „   watery, slimy, excoriating discharge: *Nat. Mur.*
   „   yellow, creamy discharge: *Nat. Phos.*

Menstruation, accompanied by:
   „   acrid leucorrhœa: *Calc. Phos., Nat. Phos.*
   „   bearing-down pains: *Calc. Fluor.*
   „   cold extremities: *Calc. Phos., Ferr. Phos.*
   „   colic: *Nat. Sulph., Kali Phos., Mag. Phos.*
   „   constipation: *Silica, Nat. Sulph.*
   „   excitableness: *Kali Phos.*
   „   flushed face: *Ferr. Phos., Calc. Phos.*
   „   fullness in abdomen: *Kali Sulph.*
   „   headache: *Nat. Mur., Ferr. Phos., Kali Phos.*
   „   hysteria: *Kali Phos.*
   „   icy coldness of body: *Silica.*
   „   labour-like pains: *Calc. Phos., Mag. Phos.*
   „   morning diarrhœa: *Nat. Sulph.*

Menstruation accompanied by:

„ nervousness: *Kali Phos.*

„ pains in back: *Calc. Phos.*

„ sadness: *Nat. Mur.*

„ watery leucorrhœa: *Nat. Mur.*

„ weeping: *Nat. Mur.*

„ weight in abdomen: *Kali Sulph.*

Menstruation, delayed, in young girls: *Nat. Mur.*

„ retarded: *Kali Mur.*

„ thin, watery blood: *Nat. Mur.*, *Kali Phos.*

„ too frequent: *Calc. Phos.*, *Kali Mur.*

„ „ late: *Nat. Mur.*, *Kali Sulph.*, *Kali Phos.*

„ „ profuse: *Kali Mur.*, *Calc. Fluor.*, *Kali Phos.*

Menstrual flow bright-red blood: *Ferr. Phos.*

„ dark, clotted, black blood: *Kali Mur.*

„ stringy and fibrous: *Mag. Phos.*

Menstruation of pale, nervous, sensitive women: *Kali Phos.*

Neuralgia of the ovaries: *Mag. Phos.*

Pains preceding monthly flow: *Mag. Phos.*

Pruritis: *Calc. Phos.*

Thighs, pain extends to: *Calc. Fluor.*

Vagina, smarting of, after urinating: *Nat. Mur.*

Vaginal secretions, acid: *Nat. Phos.*

„ „ watery, creamy, yellow: *Nat. Phos.*

RESPIRATORY ORGANS

Aching in chest: *Calc. Phos.*

Acute inflammation of the windpipe, with expectoration of frothy, watery mucus, constant frothy expectoration: *Nat. Mur.*

Acute, painful, short, irritating cough: *Ferr. Phos.*

All inflammatory conditions of the respiratory tract, in the first stage: *Ferr. Phos.*

All symptoms worse in damp weather, also rainy weather: *Nat. Sulph.*

111

Asthma, accompanied by laboured breathing: *Kali Phos.*
Asthma, aggravated by damp weather: *Nat. Sulph.*
  „ bronchial: *Kali Sulph.*
Breathing, hurried, at beginning of disease: *Ferr. Phos.*
Catch in breath: *Ferr. Phos.*
Children, cough of teething: *Calc. Phos.*
Chronic coughs: *Calc. Phos.*
Cold in chest: *Ferr. Phos.*
Constant spitting of frothy water: *Nat. Mur.*
Constriction of chest: *Mag. Phos.*
Convulsive fits of coughing: *Mag. Phos.*
Cough, better in cool open air: *Kali Sulph.*
  „ hard, dry: *Ferr. Phos.*
  „ irritating, painful: *Ferr. Phos.*
  „ pain in chest from: *Nat. Mur., Ferr. Phos.*
  „ with headache: *Nat. Mur.*
  „ „ hectic fever: *Calc. Sulph.*
  „ „ mattery expectoration: *Calc. Sulph.*
  „ worse in evening: *Kali Sulph.*
  „ „ „ warm room: *Kali Sulph.*
Croupy hoarseness: *Kali Mur., Kali Sulph.*
Expectoration, albuminous: *Calc. Phos.*
  „ difficult: *Nat. Mur., Kali Mur.*
  „ salty: *Nat. Mur.*
  „ slips back: *Kali Sulph.*
  „ streaked with blood: *Ferr. Phos.*
  „ thick, yellow, green: *Silica.*
  „ tiny yellow lumps: *Calc. Fluor., Silica.*
  „ watery: *Nat. Mur.*
  „ yellow, green, slimy: *Kali Sulph.*
Harsh breathing: *Nat. Sulph.*
Hay fever: *Mag. Phos., Nat. Mur.*
Hawking, to clear throat: *Calc. Phos.*
Hoarseness from cold: *Kali Mur.*
  „ „ over-exertion of voice: *Calc. Phos.*
  „ of speakers: *Ferr. Phos.*
Inflammatory condition of the respiratory tract, when the expectoration is decidedly yellowish, or greenish and slimy: *Kali Sulph.*

Loss of voice: *Kali Mur.*

Loud, noisy cough: *Kali Mur.*

Painful hoarseness and huskiness of speakers and singers, when due to irritating bronchi: *Ferr. Phos.*

Rattling in chest: *Kali Mur., Nat. Mur.*

Sharp pains in chest: *Mag. Phos.*

Shortness of breath from asthma or with exhaustion or want of proper nerve power; worse from motion or exertion: *Kali Phos.*

Soreness of chest: *Ferr. Phos.*

Spasmodic cough: *Mag. Phos., Kali Phos., Kali Mur.*

   ,,   ,,   at night: *Mag. Phos.*

   ,,   ,,   worse lying down: *Mag. Phos.*

Sudden, shrill voice: *Mag. Phos.*

Suffocates in heated room: *Kali Sulph.*

Stomach cough, thick, tenacious, white phlegm: *Kali Mur.*

Tickling in throat: *Calc. Fluor.*

CIRCULATORY SYMPTOMS

Anæmia: *Ferr. Phos., Calc. Phos.*

Blood, thin, watery: *Nat. Mur.*

   ,, thick, clotting: *Kali Mur.*

Blood vessels, dilatation of: *Ferr. Phos., Calc. Fluor.*

   ,,   ,, inflammation of: *Ferr. Phos.*

Circulation, poor: *Kali Phos., Calc. Phos.*

Dizziness: *Kali Phos.*

Fainting: *Kali Phos.*

   ,, from fright, grief: *Kali Phos.*

Hands and feet cold: *Calc. Phos., Nat. Mur.*

Palpitation, accompanied by nervousness: *Kali Phos.*

   ,, from indigestion: *Nat. Phos.*

   ,, inflammations: *Ferr. Phos.*

Pulse felt all over body: *Nat. Mur.*

   ,, full, rapid, quick: *Ferr. Phos.*

   ,, intermittent: *Kali Phos., Nat. Mur.*

   ,, irregular: *Kali Phos.*

   ,, sluggish: *Kali Phos., Kali Sulph.*

Pulse, subnormal: *Kali Phos.*
Relaxed veins: *Ferr. Phos., Calc. Fluor.*

BACK AND EXTREMITIES

Bunions: *Ferr. Phos., Kali Mur.*
Chilblains on hands and feet: *Kali Phos., Kali Mur.*
Coldness, feeling of, in back: *Nat. Mur.*
 „ of limbs: *Calc. Phos.*
Cracking of joints: *Nat. Phos.*
Cysts: *Calc. Phos.*
Enlargement of joints: *Calc. Fluor.*
Fingers painful or inflamed through rheumatism or
                other causes: *Ferr. Phos.*
Gout: *Nat. Phos., Nat. Sulph.*
Hang-nails: *Nat. Mur., Silica.*
Lumbago: *Calc Phos., Ferr. Phos.*
 „ from strains: *Calc. Fluor.*
 „ with dragging pain: *Calc. Fluor.*
Nails brittle: *Silica.*
Neck, muscles stiff: *Ferr. Phos.*
Numbness, feeling of: *Calc. Phos., Kali Phos.*
Rheumatism, acute: *Ferr. Phos, Nat. Phos., Nat. Sulph.,
                Silica.*
 „ with swelling: *Kali Mur.*
Rickets: *Calc. Phos.*
Sciatica: *Mag. Phos.*
Sprains: *Ferr. Phos.*
Stiff back: *Ferr. Phos.*
 „ „ worse from motion: *Ferr. Phos.*
Strains: *Ferr. Phos.*
Trembling and involuntary motion of the hands:
                Mag. Phos.
Varicose ulcerations: *Calc. Fluor.*

NERVOUS SYMPTOMS

Coldness, after attack of nervousness: *Kali Phos.*
Cramps in limbs: *Mag. Phos.*
 „ worse at night: *Calc. Phos.*

Creeping numbness: *Calc. Phos.*
Cries easily: *Kali Phos.*
Despondent: *Kali Phos.*
Dwells upon grievances: *Kali Phos.*
Feels pain keenly: *Kali Phos.*
Feet twitch during sleep: *Nat. Sulph.*
Hands twitch during sleep: *Nat. Sulph.*
Head, involuntary shaking of: *Mag. Phos.*
Hysteria: *Nat. Mur., Kali Phos.*
Involuntary motion of hands: *Mag. Phos.*
Nervous sensitiveness: *Kali Phos.*
Neuralgia accompanied by:
  ,,   congestion, after taking cold: *Ferr. Phos.*
  ,,   depression: *Kali Phos.*
  ,,   failure of strength: *Kali Phos.*
  ,,   flow of saliva: *Nat. Mur.*
  ,,   ,, ,, tears: *Nat. Mur.*
  ,,   shifting pains: *Kali Sulph.*
  ,,   ,, ,, in any organ: *Kali Phos.*
Neuralgia, obstinate, heat or cold gives no relief: *Silica.*
  ,,   ,, occurring at night: *Silica, Calc. Phos.*
  ,,   periodic: *Mag. Phos., Nat. Mur.*
  ,,   relieved by gentle motion: *Kali Phos.*
  ,,   ,, ,, pleasant excitement: *Kali Phos.*
  ,,   sensitive to light and noise: *Kali Phos.*
  ,,   worse at night: *Kali Phos.*
  ,,   ,, in cold weather: *Nat. Mur.*
  ,,   ,, ,, the morning: *Nat. Mur.*
  ,,   ,, when alone: *Kali Phos.*
Pains, like electrical shocks: *Calc. Phos., Mag. Phos.*
  ,,   ,, trickling of cold water: *Calc. Phos.*
Spasms occurring at night: *Silica, Mag. Phos., Calc. Phos.*
Trembling hands: *Mag. Phos.*

SKIN

Abscess, for heat and pain: *Ferr. Phos.*
Blisters, with fetid, watery contents: *Kali Phos.*

Blisters, with clear watery contents: *Nat. Mur.*
Burns: *Kali Mur.*
   ,,   when suppurating: *Calc. Sulph.*
Burning, as from nettles: *Calc. Phos.*
Chafed skin of infants: *Nat. Phos., Nat. Mur.*
Chapped hands from cold: *Ferr. Phos., Calc. Fluor.*
Chilblains: *Kali Phos., Kali Mur.*
Colourless, watery vesicles: *Nat. Mur.*
Cracks in palms of hands: *Calc. Fluor.*
Dandruff: *Kali Sulph., Nat. Mur., Kali Mur.*
Discharge, albuminous: *Calc. Phos.*
   ,,   blood and pus: *Calc. Sulph.*
   ,,   fetid: *Kali Phos.*
   ,,   thick, yellow pus: *Silica.*
Dry skin: *Calc. Phos., Kali Sulph.*
Eruptions with watery contents: *Nat. Mur.*
   ,,   thick, white contents: *Kali Mur.*
Excessive dryness of skin: *Nat. Mur.*
Exudations, when white and fibrinous: *Kali Mur.*
   ,,   albuminous: *Calc. Phos.*
   ,,   yellow, with small, tough lumps: *Calc. Fluor.*
   ,,   ,,   like gold: *Nat. Phos.*
   ,,   yellowish and slimy or watery: *Kali Sulph.*
   ,,   greenish, thin: *Kali Sulph.*
   ,,   clear, transparent, thin like water: *Nat. Mur.*
   ,,   mattery, or streaked with blood: *Calc. Sulph.*
   ,,   pus is thick, yellow: *Silica.*
   ,,   very offensive smelling: *Kali Phos.*
   ,,   causing soreness and chafing: *Nat. Mur., Kali Phos.*
Face full of pimples: *Calc. Phos., Calc. Sulph.*
Festers easily: *Calc. Sulph., Silica.*
Fevers, with skin dry and hot: *Kali Sulph.*
Freckles: *Calc. Phos.*
Greasy scales on skin: *Kali Phos.*
Hard, callous skin: *Calc. Fluor.*
Heals slowly: *Silica.*
Herpetic eruptions: *Nat. Mur.*

Horny skin: *Calc. Fluor.*
Inflammation of skin, for fever and heat: *Ferr. Phos.*
   „ with yellow, watery exudation: *Nat. Sulph.*
Irritating secretions: *Kali Phos.*
Irritation of the skin similar to chilblains: *Kali Mur.*
Itching, as from nettles: *Calc. Phos.*
   „ of skin, with crawling: *Kali Phos.*, *Calc. Phos.*
   „ without eruptions: *Calc. Phos.*
Ivy poisoning: *Kali Sulph.*, *Nat. Mur.*
Mattery scabs on heads of pimples: *Calc. Sulph.*
Moist scabs on skin: *Nat. Sulph.*
Nettle-rash, after becoming overheated: *Nat. Mur.*
Perspiration, lack of: *Kali Sulph.*
   „ to promote: *Kali Sulph.*
Pimples all over body, like flea-bites: *Nat. Phos.*
   „ with itching: *Calc. Phos.*
   „ under beard: *Calc. Sulph.*
Pustules on face: *Silica*, *Kali Mur.*
Rawness of skin in little children: *Nat. Phos.*
Scalds, when suppurating: *Calc. Sulph.*
Scaling eruptions on skin: *Calc. Phos.*, *Kali Sulph.*
Secretions irritate: *Kali Phos.*
Shingles: *Kali Mur.*, *Nat. Mur.*
   „ nervous symptoms: *Kali Phos.*
   „ for the pain: *Ferr. Phos.* (powder applied locally).
Skin affections with vesicular eruptions containing yellowish water: *Nat. Sulph.*
   „ yellow scabs: *Calc. Sulph.*
   „ festers easily: *Calc. Sulph.*
   „ hard and horny: *Calc. Fluor.*
   „ heals slowly and suppurates easily after injuries: *Silica.*
   „ dry, hot and burning, lack of perspiration: *Kali Sulph.*
   „ itching and burning, as from nettles: *Calc. Phos.*
   „ scales freely on a sticky base: *Kali Sulph.*
   „ withered and wrinkled: *Kali Phos.*
Suppurates easily: *Silica.*

Stings of insects: *Nat. Mur.* (applied locally).
To aid desquamation in eruptive diseases: *Kali Sulph.*
To assist in the formation of new skin: *Kali Sulph.*
Ulcers around nails: *Silica.*
Unhealthy-looking skin: *Silica.*
Yellow scabs: *Calc. Sulph.*
  „   scales on skin: *Nat. Sulph.*
Warts: *Kali Mur.*
  „   in palms of hand: *Nat. Mur.*
Wounds do not heal readily: *Calc. Sulph.*
  „   neglected, discharge pus: *Calc. Sulph.*
Wrinkled skin: *Kali Phos.*

TISSUES

Anæmia: *Calc. Phos., Ferr. Phos.*
Burns: *Kali Mur., Calc. Sulph.*
Elastic tissues relaxed: *Calc. Fluor.*
Glands, enlargement of: *Kali Mur.*
Infiltrations: *Nat. Sulph.*
Mumps: *Ferr. Phos., Kali Mur., Nat. Mur.*
Neuralgic pains in any tissue: *Mag. Phos.*
Polypi: *Calc. Phos.*
Sprains: *Ferr. Phos.*
Strains: *Ferr. Phos.*

FEVERS

Acid symptoms during fever: *Nat. Phos.*
Bilious fevers: *Nat. Sulph.*
Catarrhal fever, chilly sensations: *Ferr. Phos.*
  „   „   quickened pulse: *Ferr. Phos.*
Chill commencing in the morning about 10 o'clock and
              continuing till noon, preceded by intense
              heat, increased headache and thirst;
              sweat; great languor, emaciation, sallow
              complexion and blisters on the lips: *Nat.
              Mur.*

Chilliness at beginning of fevers: *Ferr. Phos.*
   „ in back: *Nat. Mur.*
Chills run up and down spine: *Mag. Phos.*
Clammy sweat on body: *Calc. Phos.*
Cold sweat on face: *Calc. Phos.*
Dull, heavy headache: *Nat. Mur.*
Excessive exhausting perspiration, or sweating, while eating, weakness at the stomach: *Kali Phos.*
Feeling of chilliness, especially in the back; watery saliva; full heavy headache, increased heat: *Nat. Mur.*
Fevers, during suppurative processes: *Silica.*
   „ vomit of sour fluids during: *Nat. Phos.*
   „ with chills and cramps: *Mag. Phos., Ferr. Phos.*
First stage of fevers: *Ferr. Phos.*
Flushes of heat from indigestion: *Nat. Phos.*
Frontal headache from flushes of heat: *Nat. Phos.*
Gastric fever, first stage: *Ferr. Phos.*
   „ „ when the temperature rises in the evening: *Kali Sulph.*
Increased thirst: *Nat. Mur.*
Inflammations, first stage: *Ferr. Phos.*
   „ second stage: *Kali Mur.*
In eruptive fevers to aid desquamation: *Kali Sulph.*
Much sweat in the daytime: *Nat. Mur.*
Nervous chills, with chattering of teeth: *Mag. Phos., Kali Phos.*
Nervous fever: *Kali Phos.*
Perspiration, excessive: *Calc. Phos., Kali Phos.*
   „ sour-smelling: *Nat. Phos.*
Profuse night-sweats: *Nat. Mur., Silica, Calc. Phos.*
Pulse, subnormal: *Kali Phos.*
Saliva clear, watery: *Nat. Mur.*
Shivering at beginning of fever: *Calc. Phos., Ferr. Phos.*
Sleeplessness: *Kali Phos.*
Stupor: *Nat. Mur., Kali Phos.*
To assist in promoting perspiration: *Kali Sulph.*

Tongue coated dirty, greenish-brown: *Nat. Sulph.*
  „ greyish-white, slimy: *Kali Mur.*
Twitching: *Nat. Mur.*

SLEEP

Better in evening: *Nat. Sulph.*
Constant desire to sleep in morning: *Nat. Mur.*
Drawing pain in the back at night during sleep: *Nat. Mur.*
Dreams much: *Nat. Sulph.*
  „ vivid: *Calc. Phos.*
  „ anxious: *Nat. Sulph.*
Drowsiness, with bilious symptoms: *Nat. Sulph.*
Frequent dreams and exclamations during sleep: *Silica.*
Great drowsiness: *Silica.*
Grits teeth: *Nat. Phos.*
Hard to wake in morning: *Calc. Phos.*
Heavy, anxious dreams: *Nat. Sulph.*
Jerking of limbs during sleep: *Silica, Nat. Sulph.*
Much yawning: *Silica.*
Nightmare, with bilious symptoms: *Nat. Sulph., Kali Sulph.*
Restless sleep, from worms: *Calc. Phos., Nat. Phos.*
Screams in sleep: *Nat. Phos.*
Sleep does not refresh: *Nat. Mur.*
Sleeplessness, after excitement: *Ferr. Phos., Nat. Phos.*
  „ from nervous causes: *Kali Phos.*
  „ „ worry: *Ferr. Phos., Kali Phos.*
Sleepwalking: *Kali Phos.*
Sleepy in morning: *Nat. Sulph.*
Tired in morning: *Nat. Mur., Nat. Sulph.*
Wakefulness: *Kali Phos., Ferr. Phos.*

AGGRAVATIONS AND AMELIORATIONS

Symptoms, aggravated:
  „ „ by arising from sitting position: *Kali Phos.*
  „ „ at night: *Silica, Calc. Phos.*

Symptoms, aggravated:

   ,, ,, by change of weather: *Calc. Phos.*
   ,, ,, ,, chilling feet: *Silica.*
   ,, ,, ,, cold: *Nat. Mur. Calc. Phos., Mag. Phos.*
   ,, ,, ,, cold air: *Mag. Phos., Silica.*
   ,, ,, ,, damp weather: *Calc. Phos., Nat. Sulph.*
   ,, ,, ,, draughts: *Mag. Phos.*
   ,, ,, ,, eating water plants: *Nat. Sulph.*
   ,, ,, ,, exertions: *Kali Phos.*
   ,, ,, ,, eating fish: *Nat. Sulph.*
   ,, ,, ,, fatty food: *Kali Mur.*
   ,, ,, ,, getting wet: *Calc. Phos., Calc. Sulph.*
   ,, ,, ,, heated atmosphere: *Kali Sulph.*
   ,, ,, in morning: *Nat. Sulph., Nat. Mur.*
   ,, ,, ,, evening: *Kali Sulph.*
   ,, ,, ,, open air: *Silica.*
   ,, ,, by motion: *Ferr. Phos.*
   ,, ,, ,, noise: *Kali Phos., Silica.*
   ,, ,, ,, pastry: *Kali Mur.*
   ,, ,, ,, rainy weather: *Nat. Sulph.*
   ,, ,, ,, salty atmosphere: *Nat. Mur.*
   ,, ,, ,, touch: *Mag. Phos.*
   ,, ,, ,, water: *Nat. Sulph., Calc. Sulph.*

Symptoms ameliorated by:

   ,, ,, ,, bending double: *Mag. Phos.*
   ,, ,, ,, cold: *Ferr. Phos.*
   ,, ,, ,, cool air: *Kali Sulph.*
   ,, ,, ,, eating: *Kali Phos.*
   ,, ,, ,, evening: *Nat. Mur.*
   ,, ,, ,, excitement: *Kali Phos.*
   ,, ,, ,, gentle motion: *Kali Phos.*
   ,, ,, ,, heat: *Mag. Phos., Calc. Fluor.*
   ,, ,, ,, lying down: *Calc. Phos.*
   ,, ,, ,, pleasant excitement: *Kali Phos.*

DIRECTIONS

DEFINITIONS: "Acute" means of sudden onset with more or less severity of symptoms. "Chronic" means of long continuance, lingering. A "chronic" ailment may have "acute" phases and for such phases the directions for acute conditions will apply.

Dose

In general an adult dose is four tablets, children two tablets, infants one tablet. The tablets should be dissolved on the tongue, or, in the case of very young children, may be given dissolved in a little warm water.

The Biochemic Remedies are perfectly wholesome and may be taken freely at any time. Nothing is to be gained by increasing the *size* of the dose but *frequency* of dose may be varied as occasion requires. If you discover a time for taking and frequency of dose best suited to your needs, this may be adopted.

Time

A dose should be taken every half-hour if the case is acute. In less urgent cases every two hours during the day and in chronic cases three times a day.

Alternate Remedies

In some cases it may be found that two remedies are equally strongly indicated and where this occurs such remedies should be taken in alternation. That is to say that if the case is acute, remedy "A" will be taken at the hour and remedy "B" at half past the hour and so on, repeating each remedy *hourly*. If the case is less acute, then the remedies may be taken in turn at the even hour. For instance, remedy "A", 1 o'clock, remedy "B", 2 o'clock, repeat remedy "A", 3 o'clock, repeat remedy "B", 4 o'clock and so on, repeating each remedy at *two-hourly* intervals. In chronic cases it is usual for the remedies to be taken one before meals and the other after meals, three times a day. If taken near mealtimes, at least a quarter of an hour should be allowed to elapse between meals and the taking of the remedy.

INTERCURRENT REMEDIES

This has reference to any supplementary remedy which, while of secondary importance, still has some bearing on a particular case. Intercurrent remedies are intended to assist the action of the principal remedy or remedies by correcting any secondary conditions which may have a retarding effect on the general treatment. Intercurrent remedies are usually given at bedtime and again on rising.

EXTERNAL APPLICATIONS

These are purely supplementary measures and should always be accompanied by internal treatment with similar remedies. For instance, when treating insect bites with an outward application of *Nat. Mur.*, this tissue-salt should also be given internally.

For dry application a few tablets may be crushed and the powder applied direct to the affected part. This is the usual method of dealing with cuts and abrasions (*Ferr. Phos.*), after the injured part has been cleansed.

To prepare a lotion, dissolve tablets in half a tumbler of water, which has previously been boiled and allowed to cool. This lotion may be dabbed on the affected part, such as a bruise (*Ferr. Phos.*), or a burn (*Kali Mur.*), or may be used as a cold compress, the dressing being covered with a piece of oiled silk in order to keep it moist.

A salve may be made by crushing ten tablets and mixing the powder very thoroughly with a teaspoonful of petroleum jelly, using the point of a knife, or other flat instrument, which has previously been immersed in boiling water.

POTENCIES

Biochemic Tissue-Salts should be prepared by a process of trituration in accordance with homoeopathic practice. In the preparation of each batch of tiny tablets the processes involved occupy no less than *twelve working days*. These prolonged processes are absolutely essential in order to secure the full biochemical activity and extreme fineness of these vital cell foods, which are thereby reduced practically to molecular form. These "microdoses" are readily assimilated and are absorbed almost immediately to nourish the countless millions of tiny, living cells of the human system. In homoeopathic terminology such remedies are termed potencies. These potencies may be high or low, the stages being determined on the decimal scale and indicated by the letter "x" (i.e. ten). Thus we have 3x, 6x, 12x and so on. Long experience in the first-aid treatment of everyday ailments indicates that the most generally useful potency is the 6x and this is the potency usually recommended.

INDEX

125